Advice on the Department of Energy's CLEANUP TECHNOLOGY ROADMAP

Gaps and Bridges

Committee on Development and Implementation
of a Cleanup Technology Roadmap

Nuclear and Radiation Studies Board
Division of Earth and Life Studies

NATIONAL RESEARCH COUNCIL
OF THE NATIONAL ACADEMIES

THE NATIONAL ACADEMIES PRESS
Washington, D.C.
www.nap.edu

THE NATIONAL ACADEMIES PRESS 500 Fifth Street, N.W. Washington, DC 20001

NOTICE: The project that is the subject of this report was approved by the Governing Board of the National Research Council, whose members are drawn from the councils of the National Academy of Sciences, the National Academy of Engineering, and the Institute of Medicine. The members of the committee responsible for the report were chosen for their special competences and with regard for appropriate balance.

This study was supported by Contract No. DE-FC01-04EW07022 between the National Academy of Sciences and the U.S. Department of Energy. Any opinions, findings, conclusions, or recommendations expressed in this publication are those of the author(s) and do not necessarily reflect the views of the organizations or agencies that provided support for the project.

International Standard Book Number-13: 978-0-309-13231-2
International Standard Book Number-10: 0-309-13231-2

Additional copies of this report are available from:

The National Academies Press
500 Fifth Street, N.W.
Lockbox 285
Washington, DC 20055
(800) 624-6242
(202) 334-3313 (in the Washington metropolitan area);
http://www.nap.edu

Printed in the United States of America

THE NATIONAL ACADEMIES

Advisers to the Nation on Science, Engineering, and Medicine

The **National Academy of Sciences** is a private, nonprofit, self-perpetuating society of distinguished scholars engaged in scientific and engineering research, dedicated to the furtherance of science and technology and to their use for the general welfare. Upon the authority of the charter granted to it by the Congress in 1863, the Academy has a mandate that requires it to advise the federal government on scientific and technical matters. Dr. Ralph J. Cicerone is president of the National Academy of Sciences.

The **National Academy of Engineering** was established in 1964, under the charter of the National Academy of Sciences, as a parallel organization of outstanding engineers. It is autonomous in its administration and in the selection of its members, sharing with the National Academy of Sciences the responsibility for advising the federal government. The National Academy of Engineering also sponsors engineering programs aimed at meeting national needs, encourages education and research, and recognizes the superior achievements of engineers. Dr. Charles M. Vest is president of the National Academy of Engineering.

The **Institute of Medicine** was established in 1970 by the National Academy of Sciences to secure the services of eminent members of appropriate professions in the examination of policy matters pertaining to the health of the public. The Institute acts under the responsibility given to the National Academy of Sciences by its congressional charter to be an adviser to the federal government and, upon its own initiative, to identify issues of medical care, research, and education. Dr. Harvey V. Fineberg is president of the Institute of Medicine.

The **National Research Council** was organized by the National Academy of Sciences in 1916 to associate the broad community of science and technology with the Academy's purposes of furthering knowledge and advising the federal government. Functioning in accordance with general policies determined by the Academy, the Council has become the principal operating agency of both the National Academy of Sciences and the National Academy of Engineering in providing services to the government, the public, and the scientific and engineering communities. The Council is administered jointly by both Academies and the Institute of Medicine. Dr. Ralph J. Cicerone and Dr. Charles M. Vest are chair and vice chair, respectively, of the National Research Council.

www.national-academies.org

COMMITTEE ON DEVELOPMENT AND IMPLEMENTATION OF A CLEANUP TECHNOLOGY ROADMAP

EDWIN P. PRZYBYLOWICZ (*Chair*), Eastman Kodak Company (retired), Webster, New York
ALLEN G. CROFF (*Vice Chair*), Oak Ridge National Laboratory (retired), St. Augustine, Florida
RICHELLE M. ALLEN-KING, University of Buffalo, New York
SUE B. CLARK, Washington State University, Pullman
PATRICIA J. CULLIGAN, Columbia University, New York, New York
RACHEL J. DETWILER, Braun Intertec Corporation, Bloomington, Minnesota
THOMAS F. GESELL, Idaho State University, Pocatello
GARY P. HALADA, Stony Brook University, Stony Brook, New York
CAROLYN L. HUNTOON, Consultant, Barrington, Rhode Island
EDWARD LAHODA, Westinghouse Science and Technology Center, Pittsburgh, Pennsylvania
ROBIN ROGERS, University of Alabama, Tuscaloosa
GARY S. SAYLER, University of Tennessee, Knoxville
ANDREW M. SESSLER, Lawrence Berkeley National Laboratory, Berkeley, California
J. LESLIE SMITH, University of British Columbia, Vancouver

Staff

KEVIN D. CROWLEY, Senior Board Director, Nuclear and Radiation Studies Board
JOHN WILEY, Senior Program Officer
TONI GREENLEAF, Administrative and Financial Associate
MANDI M. BOYKIN, Senior Program Assistant
SHAUNTEÉ WHETSTONE, Senior Program Assistant
JAMES YATES, JR., Office Assistant

NUCLEAR AND RADIATION STUDIES BOARD

RICHARD A. MESERVE (*Chair*), Carnegie Institution, Washington, D.C.
S. JAMES ADELSTEIN (*Vice Chair*), Harvard Medical School, Boston, Massachusetts
JOONHONG AHN, University of California, Berkeley
JOEL S. BEDFORD, Colorado State University, Fort Collins
SUE B. CLARK, Washington State University, Pullman
ALLEN G. CROFF, Oak Ridge National Laboratory (retired), St. Augustine, Florida
PATRICIA J. CULLIGAN, Columbia University, New York, New York
SARAH C. DARBY, Clinical Trial Service Unit, Oxford, United Kingdom
JAY DAVIS, Lawrence Livermore National Laboratory (retired)
ROGER L. HAGENGRUBER, University of New Mexico, Albuquerque
DAVID G. HOEL, Medical University of South Carolina, Charleston
HEDVIG HRICAK, Memorial Sloan-Kettering Cancer Center, New York, New York
THOMAS H. ISAACS, Lawrence Livermore National Laboratory, Livermore, California
PAUL A. LOCKE, Johns Hopkins Bloomberg School of Public Health, Baltimore, Maryland
BORIS F. MYASOEDOV, Russian Academy of Sciences
JOHN C. VILLFORTH, Food and Drug Law Institute (retired), Gaithersburg, Maryland
RAYMOND G. WYMER, Oak Ridge National Laboratory (retired), Oak Ridge, Tennessee
PAUL L. ZIEMER, Purdue University (retired), West Lafayette, Indiana

Staff

KEVIN D. CROWLEY, Director
MICAH D. LOWENTHAL, Senior Program Officer
JOHN R. WILEY, Senior Program Officer
DANIELA STRICKLIN, Program Officer
TONI GREENLEAF, Administrative and Financial Associate
LAURA D. LLANOS, Administrative and Financial Associate
MANDI M. BOYKIN, Senior Program Assistant
SHAUNTEÉ WHETSTONE, Senior Program Assistant
JAMES YATES, JR., Office Assistant

Preface

This study follows a series of studies by the National Research Council of the National Academies on various aspects of the cleanup of sites that formerly produced materials for this nation's nuclear defense. The Department of Energy (DOE) has requested and funded these studies over the years since the end of the Cold War. This specific study was directed at supporting the planning that is taking place within DOE's Office of Environmental Management to address some of the more challenging technical issues that are facing the cleanup task, which is expected to continue for some 30 years.

In a study such as this, the cooperation of those directly involved in the study as well as supporting agencies is paramount. The study committee found this cooperation to be outstanding. Mark Gilbertson, DOE Deputy Assistant Secretary for Engineering and Technology, and our primary customer for this study, found time in his busy schedule to attend our information-gathering meetings, and he was available for presentations and discussion at these meetings, as well as in follow-up contacts with the National Academies' staff. Beyond that, the DOE and DOE contractor contacts at each of the sites were exemplary in providing the information the committee requested. If this report lacks certain details, it is an oversight by the committee in not requesting such information.

The study and report content were strongly aided and abetted by expertise at the National Academies in the persons of Kevin Crowley and John Wiley. Both have had extensive experience in leading studies specifically in this field of waste cleanup, and also in related fields. Their input relative to pertinent background material, reports, and contacts was invaluable.

Finally, John Wiley is especially recognized for his broad skills as a Senior Program Officer at the National Academies for doing all that is expected of a program director, from structuring meetings, arranging for presentations to the committee, arranging site visits, and steering the committee to see the right things on these visits to providing a writing capability in what he prefers to refer to as "diddling" and "mangling," that has made this report not only reflect the committee's observations, findings, and recommendations, but doing so in clear and understandable fashion. The committee owes John a debt of gratitude for his support and commitment throughout this study.

Backing up Kevin Crowley and John Wiley at the National Academies are administrative assistants who should be recognized for their contributions in supporting the committee in a variety of ways from travel arrangements, website portal support, reference reports, telephone conferences, formatting reports and, despite the "paperless society" of today, spending endless time at the copying machine to provide materials for meetings. Our grateful thanks to Mandi Boykin, who supported this committee through most of its study and meetings; Toni Greenleaf, who monitored our budget and provided helpful backup support for Mandi; and Shaunteé Whetstone, who helped us in the final stages of producing this report.

On a personal note, the Chair would like to thank Vice Chairman Allen Croff whose expertise in many elements of this study are based not only on his career at the Oak Ridge National Laboratory and as an advisor to the U.S. Nuclear Regulatory Commission but also his service on many National Academies committees. Allen has provided wise counsel in the course of this study and, in a couple of critical meetings that the Chair was not able to attend, stepped in to effectively conduct the proceedings.

Edwin Przybylowicz, Chair

Allen Croff, Vice Chair

Reviewer Acknowledgments

This report has been reviewed in draft form by individuals chosen for their diverse perspectives and technical expertise in accordance with procedures approved by the National Research Council's Report Review Committee. The purpose of this independent review is to provide candid and critical comments that will assist the institution in making its published report as sound as possible and to ensure that the report meets institutional standards of objectivity, evidence, and responsiveness to the study charge. The content of the review comments and draft manuscript remain confidential to protect the integrity of the deliberative process. We wish to thank the following individuals for their participation in the review of this report:

Rudolph Bonaparte, Geosyntec Consultants, Atlanta, Georgia
Ken Czerwinski, University of Nevada, Las Vegas
Harry Harmon, Consultant, Aiken, South Carolina
Paul Locke, Johns Hopkins University, Baltimore, Maryland
Bruce Moyer, Oak Ridge National Laboratory, Oak Ridge, Tennessee
Walter Robb, Vantage Management, Inc., Schenectady, New York
Ray Wymer, Oak Ridge National Laboratory (retired), Oak Ridge, Tennessee
John Zachara, Pacific Northwest National Laboratory, Richland, Washington

Although the reviewers listed above have provided many constructive comments and suggestions, they were not asked to endorse the report's conclusions or recommendations, nor did they see the final draft of the report before its release. The review of this report was overseen by Chris G. Whipple, ENVIRON International Corporation. Appointed by the Division on Earth and Life Sciences, he was responsible for making certain that an independent examination of this report was carried out in accordance with institutional procedures and that all review comments were carefully considered. Responsibility for the final content of this report rests entirely with the authoring committee and the National Research Council.

Contents

Overview

The Department of Energy's (DOE's) fiscal year 2009 budget request put the potential cost of removing or remediating radioactive waste and other contamination at its former nuclear weapons production sites between $265 billion and $305 billion over the next approximately 30 years.[1] DOE has stated that this work, which is being conducted by its Office of Environmental Management (EM), represents one of the most technically challenging and complex cleanup efforts in the world. Furthermore, DOE noted that the future course of the Department's environmental cleanup activities will depend on a number of fundamental technical and policy choices, many of which have not been made (DOE 2008a, p. 16).

To enhance its cleanup efforts, EM has invested in scientific research and technology development. The funding for these investments has been inconsistent and generally decreasing from a peak of almost $410 million in fiscal year 1995 to around $20 million per year recently—about 0.4 percent of EM's overall budget. There has, however, been renewed interest in cleanup science and technology development, both within upper DOE management and in Congress. In early 2007, EM turned to the National Academies for assistance in preparing a congressionally requested engineering and technology roadmap to support the cleanup effort.

The statement of task for this study directed the committee to identify (1) principal science and technology gaps and their priorities for the cleanup program, (2) expertise and infrastructure at the national laboratories that should be maintained to address the higher priority cleanup challenges, and

[1] See http://www.cfo.doe.gov/budget/09budget/Content/Volumes/Volume5.pdf.

(3) strategic opportunities to leverage research and development (R&D) with other organizations. The committee was asked to focus on the DOE's four major cleanup sites: the Hanford Reservation, Washington; the Idaho National Laboratory; the Oak Ridge Reservation, Tennessee; and the Savannah River Site, South Carolina.

The committee chose as its working definition that *a gap is a shortfall in available knowledge or technology that could prevent EM from accomplishing a cleanup task on its expected schedule and/or budget.* Using this definition, the committee identified and detailed 13 gaps in areas of tank waste retrieval and processing, groundwater and soil remediation, and facility deactivation and decommissioning that could adversely affect EM's ability to meet its cleanup milestones on time and/or on budget. In order to conduct R&D toward bridging these gaps, the sites and national laboratories will need to maintain certain critical expertise and infrastructure for:

- Handling radioactive materials,
- Conducting engineering and pilot-scale tests,
- Determining contaminant behavior in the environment, and
- Utilizing relevant state-of-the-art science to develop advanced cleanup technologies.

EM's Office of Engineering and Technology should partner with other DOE offices, other federal agencies, academia, and the private sector in order to provide the needed science and technology for advanced cleanup methodologies. Partnering with these other resources can provide the lowest cost means to address technology gaps in EM's roadmap in two important ways: (1) it takes advantage of science and technology relevant to its cleanup task that is being developed in other laboratories throughout the world, and (2) it keeps to a minimum the R&D for which EM has to provide direct and total support. EM can bring to its partnerships unique onsite facilities, detailed data on its groundwater and soil contamination, and decades of experience in managing radioactive wastes.

The committee provided findings and recommendations in two areas: (1) improving the EM roadmap so that it clearly details the role of R&D in the EM cleanup mission, and (2) roadmapping R&D programs that utilize national laboratory, site, and private-sector capabilities to bridge the science and technology gaps identified by the committee. This report concludes with the committee's observations on how EM's Office of Engineering and Technology can enhance its role in leading EM's R&D programs.

Summary

Beginning with the Manhattan Project and continuing through the Cold War, the U.S. government constructed and operated a massive industrial complex to produce and test nuclear weapons and related technologies. When the Cold War ended, most of this complex was shut down permanently or placed on standby, and the government began a costly, long-term effort to clean up the wastes and environmental contamination resulting from its nuclear materials production.

In 1989, Congress created the Office of Environmental Management (EM) within the Department of Energy (DOE) to manage this cleanup effort. EM has made substantial progress—for example, decommissioning of the Rocky Flats Site, perhaps the nation's most highly contaminated plutonium facility, in 2005, ahead of schedule and under budget. The site became a national wildlife refuge in 2007. EM has also completed other significant site cleanups.[1] Nonetheless, the scope of EM's remaining future cleanup work is enormous.

DOE's fiscal year 2009 budget request put the potential cost of removing or remediating radioactive waste and other contamination at the sites between \$265 billion and \$305 billion over the next approximately 30 years.[2] DOE has stated that the EM cleanup represents one of the most technically challenging and complex cleanup efforts in the world, and furthermore that the future course of the Department's environmental cleanup

[1] See http://www.em.doe.gov/Pages/BudgetPerformance.aspx.

[2] See http://www.cfo.doe.gov/budget/09budget/Content/Volumes/Volume5.pdf.

activities will depend on a number of fundamental technical and policy choices, many of which have yet to be made (DOE 2008a, p. 16).

To enhance its cleanup efforts, EM has invested in scientific research and technology development. The funding for these investments has been inconsistent and generally decreasing from a peak of almost $410 million in fiscal year 1995 to around $20 million per year recently.[3] There has, however, been renewed interest in cleanup science and technology development, both within upper DOE management and in Congress, as well as among citizens who reside near the sites. The fiscal year 2007 House Energy and Water Development Appropriations Report[4] requested that EM provide an engineering and technology roadmap to help justify future, sustained R&D support. The roadmap was to identify technology gaps in the current DOE site cleanup program and give a plan to address them. For assistance, EM's Office of Engineering and Technology (EM-20) turned to the National Research Council of the National Academies, which empaneled a committee to undertake the study that it has described in this report. The committee carried out its task with the intent of assisting and strengthening EM's roadmapping efforts.

THE COMMITTEE'S TASK

The technology roadmapping process has been widely used as a planning tool in industry and government to match technology resources with desired product or process outputs. To assist EM with its Roadmap, the statement of task for this study (Chapter 1, Sidebar 1.2) directed the committee to identify (1) principal science and technology gaps and their priorities for the cleanup program; (2) strategic opportunities to leverage needed research and development programs from other DOE programs, federal agencies, universities, and the private sector; (3) core capabilities at the national laboratories that will be needed to address EM's long-term, high-risk cleanup challenges; and (4) infrastructure at national laboratories and EM sites that should be maintained to support research, development, and demonstrations of cleanup technologies. The committee was asked to focus on the DOE's four major cleanup sites—the Hanford Reservation, Washington; the Idaho National Laboratory (INL); the Oak Ridge Reservation (OR), Tennessee; and the Savannah River Site (SRS), South Carolina—and

[3] EM's appropriation for technology development and deployment (EM-20) was just over $21 million, about 0.4 percent of the total EM appropriation of about $5.7 billion in fiscal year 2008. The fiscal year 2009 request for EM-20 is just over $32 million. See http://www.em.doe.gov/Pages/budgetdocs.aspx.

[4] House Report 109-474 to accompany H.R. 5427, Energy and Water Development Appropriations Bill, 2007.

to provide findings and recommendations on maintenance of core capabilities and infrastructure at the national laboratories and the sites.

The committee began its study with a workshop in March 2007 at which it heard from DOE headquarters and site representatives, regulators, and citizens who described cleanup challenges and technology needs at Hanford, INL, OR, and SRS. The committee then visited these four sites and their associated national laboratories. An interim report was released in February 2008 to assist EM with its fiscal year 2009 planning (NRC 2008). Three observations from the interim report served to guide the committee's subsequent deliberations that led to this report:

Observation 1: The complexity and enormity of EM's cleanup task require the results from a significant, ongoing R&D program so that EM can complete its cleanup mission safely, cost-effectively, and expeditiously.

Observation 2: By identifying the highest cost and/or risk aspects of the site cleanup program, the EM roadmap can be an important tool for guiding DOE headquarters investments in longer-term R&D to support efficient and safe cleanup.

Observation 3: The national laboratories at each site have special capabilities and infrastructure in science and technology that are needed to address EM's longer-term site cleanup needs. The EM roadmap can help establish a more direct coupling of the national laboratories' capabilities and infrastructures with EM's needs.

The committee's final information-gathering meeting, in April 2008, addressed opportunities for EM to leverage its research with other organizations.

SCIENCE AND TECHNOLOGY GAPS

To address its statement of task in a way it judged would be most useful to EM, the committee chose as its working definition that *a gap is a shortfall in available knowledge or technology that could prevent EM from accomplishing a cleanup task on its expected schedule and/or budget.* Following the analogy of a roadmap, a science and technology gap is a pothole in the road that EM might somehow work around, but at the likely cost of time and money. It would be much better to fill the pothole or avoid it altogether with appropriate R&D.

The committee used the major program areas in EM's draft Engineering and Technology Roadmap (DOE 2007a) to frame its gap identification, although it used its own deliberations and judgment to identify technology gaps. Chapter 2 of this report gives an assessment of each of these gaps: its context in the EM cleanup, its potential impacts, relevant work in progress, and suggestions for R&D to help bridge the gap. Each assessment also summarizes the committee's rationale for prioritizing each gap, as required by the task statement. However, all gaps that the committee chose to describe in this report have the potential to adversely affect EM's ability to meet its cleanup milestones on time and/or on budget. The prioritization is therefore relative among gaps within each program area and all are significant enough to be roadmapped for R&D. For more detail regarding the committee's gap analyses, see Chapter 2.

Roadmap Program Area: Waste Processing

- Substantial amounts of waste may be left in tanks/bins after their cleanout—especially in tanks with obstructions, compromised integrity, or associated piping (Priority: High).
- Increased vitrification capacity may be needed to meet schedule requirements of EM's high-level waste programs (High).
- Low-activity streams from tank waste processing could contain substantial amounts of radionuclides (Medium).
- New facility designs, processes, and operations usually rely on pilot-scale testing with simulated rather than actual wastes (Medium).
- The baseline tank waste vitrification process significantly increases the volume of high-level waste to be disposed (Medium).
- A variety of wastes and nuclear materials do not yet have a disposition path (Low).

Roadmap Program Area: Groundwater and Soil Remediation

- The behavior of contaminants in the subsurface is poorly understood (High).
- The long-term ability of cementitious materials to isolate wastes is not demonstrated (High).
- Site and contaminant source characteristics may limit the usefulness of EM's baseline subsurface remediation technologies (Medium).
- The long-term performance of trench caps, liners, and reactive barriers cannot be assessed with current knowledge (Medium).

Roadmap Program Area: Facility Deactivation and Decommissioning (D&D)

- D&D work relies on manual labor for building characterization, equipment removal, and dismantlement (High).

- Removing contamination from building walls, other surfaces, and equipment can be slow and ineffective (Medium).
- Personal protective equipment tends to be heavy, hot, and limits movement of workers (Low).

EXPERTISE AND INFRASTRUCTURE AT THE NATIONAL LABORATORIES AND EM SITES

After reviewing the science and technology gaps identified in Chapter 2, the committee determined that in order to conduct R&D toward bridging these gaps the sites and national laboratories will need to maintain the expertise and infrastructure for:

- Handling radioactive materials,
- Conducting engineering and pilot-scale tests,
- Determining contaminant behavior in the environment, and
- Utilizing state-of-the-art science to develop advanced technologies.

These capabilities[5] are described briefly below and in detail in Chapter 3.

The capability to work with radioactive materials is fundamental to EM's engineering and technology development. All of the national laboratories visited by the committee have this capability, which includes the ability to perform chemical analyses and provide health physics support. Comparable capability does not exist outside of the national laboratories. Laboratory personnel become qualified to work with radioactive materials almost exclusively through onsite training and experience. EM's 30-year cleanup program cannot be sustained without this capability.

Engineering and technology development activities include testing to provide basic parameters to design new processes and equipment (e.g., heat and mass transfer, mixing, and corrosion) and to demonstrate them at the pilot scale or larger. Capabilities for engineering and pilot-scale testing are needed to support R&D to address most of the gaps identified in Chapter 2. The capability to conduct engineering tests is not unique to the sites and national laboratories, but there are instances for which they are best suited—for example, the high-level waste tank mock-up facilities at Hanford, INL, and SRS. Technicians and operators who have experience with the site problems that their work is addressing often contribute innovative, practical ideas for their solutions.

Each of the DOE sites has a unique history in the disposal or release of contamination and unique geohydrological characteristics, which largely

[5] "Capability" is used by the committee to refer to both personnel expertise and physical infrastructure.

control the movement of these contaminants. Contamination has reached the groundwater at all four sites visited by the committee. Groundwater and soil remediation will likely continue for the duration of the EM cleanup. Capabilities for determining contaminant behavior in the environment are needed to support R&D to address all of the groundwater and soil gaps identified in Chapter 2. Some of these capabilities are unique to the sites and their associated national laboratories, including groundwater sampling facilities, experimental barriers against contaminant migration, and the accumulated knowledge of site history and geohydrology among long-term employees.

Presentations by the national laboratories during the committee's site visits and by DOE's Office of Science during the committee's April 2008 meeting provided an overview of many advanced scientific capabilities applicable to addressing the science and technology gaps identified in Chapter 2. Furthermore, it is clear that the state-of-the-art science and technology relevant to EM's cleanup task will advance over the next 30 years of the EM cleanup in ways that can only be imagined today. While EM would not be expected to be a primary user or primary financial supporter of advanced scientific facilities, it is essential that EM and the Office of Science continue close cooperation and coordination to ensure that EM is able to utilize state-of-the-art science and that the national laboratories put effort into solving EM's unique problems.

LEVERAGING ENVIRONMENTAL MANAGEMENT R&D WITH OTHER ORGANIZATIONS

The committee paid special attention to this part of its task statement because EM's Office of Engineering and Technology does not have the resources necessary to sustain all of the capabilities necessary for its R&D work, which are described in Chapter 3. As a consequence, a large portion of the R&D work that is needed by EM will involve partnering with other organizations.

The committee judged that the effectiveness of EM's efforts to leverage its R&D investments can be enhanced by:

1. Improving the Roadmap and using it as a central tool for EM's R&D planning and for communicating its plans and programs to other organizations, including other DOE offices, federal agencies, and Congress, and
2. Better application of the basic principles of leveraging research, with recognition that legacy waste cleanup is a national responsibility that requires other organizations to partner willingly with EM.

To be successful in leveraging R&D, all participants in the collaboration must receive benefits from the partnership in order for it to be sustained. Moreover, all participants in the collaboration should bring something to the partnership that is needed by the other partners in the collaboration. This may range from financial resources to specific capabilities that other partners can build on and benefit from.

In the planning and development of its Roadmap, details, time lines, and close interactions with potential leveraging partners can help ensure that there are viable connections between EM's roadmapped objectives and the support it can negotiate with these partners. Notably the 2008 EM roadmap provides no time lines for its initiatives or connections between the initiatives and EM site cleanup milestones. This is rather like drawing a map by simply listing cities without placing them geographically on the map or showing highway interconnections. A much more useful EM roadmap will show when and how the initiatives address technology gaps such as those identified in this study.

In identifying partnership opportunities, which can result in true leveraging among the participant organizations (i.e., both organizations benefiting from the relationship), the committee wishes to reemphasize the necessary *quid pro quo* nature of these partnerships and the need to ensure that EM is fully vested to enter into such relationships as an equal partner.

FINDINGS AND RECOMMENDATIONS

The statement of task directed the committee to provide findings and recommendations, as appropriate, to EM on maintenance of core capabilities and infrastructure at national laboratories and EM sites to address its long-term, high-risk cleanup challenges. In carrying out its task, the committee judged that EM's Engineering and Technology Roadmap can be a key tool to ensure that core capabilities and infrastructure remain available to EM over the next 30 years of the site cleanup program.

The committee's findings and recommendations in Chapter 5 address two topics: (1) improving the Roadmap so that it clearly demonstrates the role of R&D in the EM cleanup mission, and (2) establishing R&D programs that utilize national laboratory, site, and private-sector capabilities to bridge the science and technology gaps identified in Chapter 2. At the end of Chapter 5 the committee gives a concluding set of observations that may help strengthen future initiatives by the EM Office of Engineering and Technology to bring new technologies into the EM cleanup effort.

Improving the Roadmap

FINDING: The EM Engineering and Technology Roadmap is an important and much needed tool for guiding DOE headquarters investments in longer-term R&D to support efficient and safe cleanup.

FINDING: The current Roadmap describes technical risks in the EM site cleanup program and R&D initiatives to mitigate these risks. However, it does not connect these initiatives to major milestones in the EM cleanup program.

RECOMMENDATION 1: **EM's Office of Engineering and Technology should update its 2008 Roadmap to include performance metrics and timelines for accomplishing its R&D initiatives to ensure that results are useful and timely to meet EM's site cleanup milestones.**

RECOMMENDATION 2: **The DOE Assistant Secretary for Environmental Management should require periodic, future updates of the Roadmap to ensure that it remains current with major mid- to long-term milestones in the cleanup program. At a minimum, the Roadmap should be updated at least every 4 years at an appropriate time to help ensure carryover of programs and their rationales into new administrations.**

FINDING: EM is the DOE office designated to clean up the nuclear materials production sites of the Cold War. Cleaning up these legacy sites nevertheless remains a responsibility for all of DOE and the nation. EM cannot complete its mission without the active cooperation of other DOE offices and federal agencies. The Roadmap can be improved by specifying opportunities for cooperative work with the national laboratories and other DOE and federal agencies.

RECOMMENDATION 3: **The EM Office of Engineering and Technology, with support of the Secretary of Energy and the Assistant Secretary for Environmental Management, should engage other federal organizations (e.g., Department of Defense, Department of Homeland Security, Environmental Protection Agency) and DOE offices (e.g., Office of Science, Office of Nuclear Energy, Office of Legacy Management) to specify Roadmap intersections with the others' R&D programs to ensure that opportunities for joint work are recognized and implemented in timely fashion to produce results that are useful to EM.**

EM could do this by convening workshops at which participants exchange information on their cleanup-relevant R&D programs and milestones. The Office of Engineering and Technology did this to a limited extent in preparing the 2008 Roadmap. The workshops could be arranged to provide timely information for periodic updates of the Roadmap according to Recommendation 2.

RECOMMENDATION 4: **The DOE Assistant Secretary for Environmental Management and the Office of Engineering and Technology should use the Roadmap as a primary means of communicating EM's technology needs, R&D planning, and accomplishments within DOE, to other federal and state agencies, and ultimately to Congress.**

FINDING: The scientific and technical state of the art will evolve during the next 30 years of the EM site cleanup program, as will public expectations for the cleanup goals. A robust EM science, engineering, and technology program will be required to keep up with these evolutions, to provide up-to-date bases for EM's cleanup decisions, and to maintain a skilled workforce.

RECOMMENDATION 5: **EM and its Office of Engineering and Technology should include in its Roadmap the overarching themes of (1) maintaining state-of-the-art cleanup objectives as science, technology, and the public's expectations evolve during the next 30 years; (2) maintaining and distributing up-to-date knowledge resources relevant to site cleanup; and (3) developing a balanced R&D portfolio that addresses short-, medium-, and long-term issues.**

In the first instance, the Roadmap might identify organizations responsible for providing technical data and timely R&D milestones to support key EM site cleanup decisions (e.g., the cleanup objective for a waste burial ground, a groundwater plume, or a decommissioned facility). In the second instance, the Roadmap might include objectives for hiring and retaining personnel, and for information archiving, at specified milestone times during the next 30 years.

Bridging EM's Science and Technology Gaps

FINDING: The unique chemical, physical, and radiological properties of waste and contamination at the EM cleanup sites and the unique subsurface characteristics of the sites themselves require special capabilities of the sites and their associated national laboratories to sustain long-term R&D for EM's 30-year cleanup program. These special capabilities include qualified, experienced personnel and facilities for radiochemical, engineering, and field experiments. It is Congress's and DOE's responsibility to maintain the national laboratories' capabilities not only for cutting-edge scientific research but also for research applied to national problems such as DOE's Cold War legacy cleanup.

RECOMMENDATION 6: **The EM Office of Engineering and Technology, with support from the Secretary of Energy and the Assistant Secretary for Environmental Management, should lay out in its Roadmap programs that include research in the following:**

- **Radiochemistry of EM wastes and contaminants;**
- **Long-term performance of cementitious materials;**
- **Retrieval technology for high-level waste;**
- **Alternative and advanced waste forms and production methods;**
- **Rheology of waste sludges and slurries;**
- **Long-term behavior of in-ground contaminants;**
- **Advanced sensors, detectors, and data transmission technology for subsurface monitoring;**
- **Advanced near-surface engineered barrier systems to control contaminant release to the environment; and**
- **Surface characterization of solid materials.**

Each of these recommended programs is described in Chapter 5.

CONCLUSION

At the beginning of the study the committee understood that the Roadmap would be a living document to help plan, justify, and increase the effectiveness of EM's R&D program in support of its site cleanup mission. The committee found that the Roadmap can be an important tool for enhancing EM's R&D efforts and has recommended detailed improvements and periodic updates of the Roadmap. We hope that this report with its findings and recommendations will be useful to DOE and to EM.

1

Introduction

Beginning with the Manhattan Project and continuing through the Cold War, the U.S. government constructed and operated a massive industrial complex to produce and test nuclear weapons and related technologies. At its peak, this complex encompassed over 100 distinct sites in 31 states and one territory with a total area of over two million acres (DOE 1997). Most of the nuclear material production and recycling operations took place at five sites: the Hanford Reservation, Washington; the Idaho National Laboratory (INL); the Oak Ridge Reservation (OR), Tennessee; the Rocky Flats Site, Colorado; and the Savannah River Site (SRS), South Carolina. These sites supplied large quantities of nuclear materials, primarily plutonium, highly enriched uranium, and tritium.

When the Cold War ended, most of this complex was shut down permanently or placed on standby, and the U.S. government began a costly, long-term effort to clean up the materials, wastes, and environmental contamination resulting from its nuclear materials production. In 1989, Congress created the Office of Environmental Management (EM) within the Department of Energy (DOE) to manage this cleanup effort. EM has made substantial progress. Decommissioning of the Rocky Flats Site, perhaps the nation's most highly contaminated plutonium facility, was completed ahead of schedule and under budget in 2005, and the site became a national wildlife refuge in 2007. Other cleanup accomplishments are tracked on EM's website.[1]

Nonetheless, the scope of EM's future cleanup work is enormous.

[1] See http://www.em.doe.gov/Pages/BudgetPerformance.aspx.

DOE's fiscal year 2009 budget request put the potential cost of removing or remediating radioactive waste and other contamination at the sites between $265 billion and $305 billion over the next approximately 30 years.[2] This is a major liability for DOE and for the nation. According to DOE's 2008 Agency Financial Report (DOE 2008a, p. 16):

> The Department has significant unfunded liabilities that will require future appropriations to fund. The most significant of these represent ongoing efforts to clean up environmental contamination resulting from past operations of the nuclear weapons complex. The FY 2008 environmental liability estimate totaled $266 billion and *represents one of the most technically challenging and complex cleanup efforts in the world.*
>
> Estimating this liability requires making assumptions about future activities and is inherently uncertain. *The future course of the Department's environmental cleanup activities will depend on a number of fundamental technical and policy choices, many of which have not been made.* The cost and environmental implications of alternative choices can be profound. [Italics added]

From its inception, the EM program has faced three fundamental technical challenges: First, to inventory and characterize the vast array of materials, wastes, and contamination resulting from weapons production, testing, and related activities. Second, to decide whether, how much, and when to retrieve, treat, remediate, or dispose of these materials, wastes, and contamination. Third, how to implement the cleanup operations in a timely manner.

EM has made major investments in scientific research and technology development to obtain the needed knowledge and tools to meet these challenges. There has been a technology development program within EM since its creation.[3] However, headquarters-directed investments in science and technology activities have varied substantially, rising from $184 million in fiscal year 1990 to almost $410 million in fiscal year 1995, followed by a decade-long decline to $21.2 million in fiscal year 2008. This amounted to about 0.4 percent of EM's total appropriation for fiscal year 2008.[4] Beginning in about 2002, the program became focused almost exclusively on short-term technology development needs to support accelerated site cleanup (DOE 2002).

There has been recent renewed interest in longer-term cleanup science

[2] See http://www.cfo.doe.gov/budget/09budget/Content/Volumes/Volume5.pdf.

[3] This technology development program has had several names: Office of Technology Development (1990-1995); Office of Science and Technology (1995-2003); Office of Environmental Cleanup & Acceleration (2003-2006); and Office of Engineering and Technology (May 2006 to present).

[4] See http://www.em.doe.gov/Pages/budgetdocs.aspx.

and technology development, both within upper DOE management and in Congress. This interest was inspired in part by reports of the National Research Council (NRC). A congressionally mandated study (NRC 2006b) evaluated DOE's plans for retrieval and onsite disposal of certain wastes stored in tanks at the Hanford, Idaho, and Savannah River sites. The report from that study recommended (NRC 2006b, pp. 6-7) that DOE initiate a targeted, aggressive, and collaborative research and development (R&D) program to support its efforts to retrieve waste and clean and close tanks in which this waste is currently being stored. It further recommended that this R&D program last for 10 years with funding on the order of $50 million per year.

The fiscal year 2007 House Energy and Water Development Appropriations Report[5] recommended an increase in EM's R&D funding and, to sustain future support, requested that EM provide an engineering and technology roadmap. The roadmap was to identify technology gaps in the current DOE site cleanup program and a strategy, with funding proposals, to address them. The Appropriations Report cited another previous NRC report (2005), as follows:

> The EM technology development program funding has declined over the years, while at the same time, many technological challenges continue to face the program. For example, the National Research Council's 2005 report on *Improving the Characterization and Treatment of Radioactive Wastes* recommends that "an improved capability for environmental monitoring would strengthen EM's plans to leave waste and contaminated media at DOE sites," and, "Monitoring systems at EM closure sites have been estimated to be some 25 years behind the state-of-art." The Committee directs the increase to address the technology short-falls identified by this report.

As EM began work on the Roadmap, the DOE Assistant Secretary for Environmental Management and the EM Office of Engineering and Technology turned to the NRC for assistance. The NRC in turn empaneled the committee that prepared this final report. The committee held its first meeting, as a workshop, in March 2007. In April 2007, EM issued a draft of its Engineering and Technology Roadmap, which provided the basis of much of the committee's information gathering and deliberations.[6] EM issued its first Roadmap in final form in March 2008 (Sidebar 1.1).

---

[5] House Report 109-474 to accompany H.R. 5427, Energy and Water Development Appropriations Bill, 2007.

[6] EM's Engineering and Technology Roadmap is referred to as the EM roadmap or as the Roadmap throughout this report.

SIDEBAR 1.1
A Brief Description of the EM Cleanup Technology Roadmap[a]

The technology roadmapping process has been widely used as a planning tool in industry and government to match technology resources with desired product or process outputs. In the case of industry, these outputs are often products to meet certain commercialization needs. In *Vision 2020: The Lighting Technology Roadmap*, DOE used this technique in working with industry to align resources to meet new challenges in building lighting systems (DOE 2007b).

The EM roadmap lists five program areas that are central to site cleanup:

1. Tank waste processing (including waste retrieval and tank closure),
2. Groundwater and soil remediation (including buried waste, flow path, and contaminant characterization),
3. Facility deactivation and decommissioning,
4. DOE spent nuclear fuel, and
5. Challenging materials (generally speaking, these are nuclear materials with no definite path to disposition).

Technical risks and uncertainties are listed in tabular format for each of these program areas. For example, within tank waste processing, the Roadmap indicates that there are technical risks and uncertainties involving waste storage, waste retrieval, tank closure, waste pretreatment, and stabilization. Strategic initiatives to address each uncertainty are also listed.

[a]DOE (2007a, 2008b).

THE COMMITTEE'S APPROACH

The Statement of Task for this committee, the Committee for the Development and Implementation of a Cleanup Technology Roadmap,[7] asks for advice to support the development of a cleanup technology roadmap for EM (Sidebar 1.2). The committee was to identify (1) existing technology gaps and their priorities, (2) strategic opportunities to leverage needed R&D programs with other organizations, (3) needed core capabilities, and (4) infrastructure at national laboratories and EM sites that should be maintained to accomplish EM's mission.

A technology roadmap is a tool or a disciplined way to plan and couple R&D programs to needed outputs. It is a well-established planning tool in the private sector as well as in government agencies. Given the complex-

[7] Referred to as the committee throughout this report.

SIDEBAR 1.2
Statement of Task

A National Academies committee will provide technical and strategic advice to the DOE-EM's Office of Engineering and Technology to support the development and implementation of its cleanup technology roadmap. Specifically, the study will identify:

- Principal science and technology gaps and their priorities for the cleanup program based on previous National Academies' reports, updated and extended to reflect current site conditions and EM priorities and input from key external groups, such as the Nuclear Regulatory Commission, Defense Nuclear Facilities Safety Board, Environmental Protection Agency, and state regulatory agencies.
- Strategic opportunities to leverage research and development from other DOE programs (e.g., in the Office of Science, Office of Civilian Radioactive Waste Management, and the National Nuclear Security Administration), other federal agencies (e.g., Department of Defense, Environmental Protection Agency), universities, and the private sector.
- Core capabilities at the national laboratories that will be needed to address EM's long-term, high-risk cleanup challenges, especially at the four laboratories located at the large DOE sites (Idaho National Laboratory, Oak Ridge National Laboratory, Pacific Northwest National Laboratory, and Savannah River National Laboratory).
- The infrastructure at these national laboratories and at EM sites that should be maintained to support research, development, and bench- and pilot-scale demonstrations of technologies for the EM cleanup program, especially in radiochemistry.

The committee will provide findings and recommendations, as appropriate, to EM on maintenance of core capabilities and infrastructure at national laboratories and EM sites to address its long-term, high-risk cleanup challenges.

ity of EM's cleanup mission and the challenging technologies that must be invented or adapted and utilized to support that mission, a technology roadmap is well suited and indeed essential to plan the future course of EM R&D.

The committee included experts in disciplines relevant to the cleanup task, including chemistry and radiochemistry, geoscience, materials science, nuclear and chemical engineering, and health physics. In addition, management expertise in government, academia, national laboratories, and the private sector was represented (Appendix A). The balanced expertise in this committee enabled broad-ranging discussion of the issues related to supporting the development of the Roadmap.

EM's cleanup program has been the focus of numerous studies by the

National Academies over the past 15 years. This committee's work was "jump started" with a workshop in March 2007 to review the status of the cleanup efforts using the previous studies and updates from the various sites (NRC 2007c). The workshop provided an effective summary of the state of EM's cleanup program, which became the departure point for the committee's visits to the four major operating sites: the Hanford Reservation, INL, OR, and SRS. Following each of the site visits, each committee member provided factual input on what he or she had come away with from the visit. This input is summarized in the Appendixes of this report and forms the factual basis for the committee's synthesis of observations, findings, and recommendations. At the request of the committee, the Appendixes were fact checked by the sites.

At roughly the midpoint of its study and at the request of EM, the committee prepared an interim report (Appendix H) that presented three observations that highlighted the needs for an ongoing EM R&D program and longer-term support for this program. The remainder of the study helped confirm the interim report's early observations. These themes of "ongoing" and "longer-term" are carried through this final report. Sidebar 1.3 describes how the committee viewed the time frames for EM's roadmapping and R&D programs and used them in developing this report. The three observations provided a context for its final findings and recommendations in Chapter 5.

In its deliberations, the committee put together an initial matrix of all the technology gaps it had identified (Appendix C). This list included about 50 items, which were narrowed down in later discussions to yield what the committee considered to be the principal technology gaps facing EM. These are summarized in Chapter 2. The visits to the sites also provided the basis of identifying core capabilities and infrastructure that should be maintained to support EM's long-term cleanup goals (Chapter 3).

As noted earlier, the committee used the draft EM roadmap, issued in April 2007, to guide its information gathering and deliberations. The draft Roadmap included only the first three program areas listed in Sidebar 1.1. The last two areas, spent nuclear fuels and challenging materials, were added in EM's final Roadmap (DOE 2008b). They were not discussed by the committee as separate program areas. Nevertheless, technology gaps related to them were apparent from the committee's information gathering, and they are included in one of the gap analyses in Chapter 2.

After its site visits, the committee focused its last information-gathering meeting on the topic of leveraging R&D. In this meeting, the committee reviewed the processes of roadmapping and partnering as tools in planning and implementing the leveraging of EM R&D. In addition, a number of organizations from DOE and other federal agencies presented leveraging opportunities to the committee. This provided the basis for Chapter 4.

SIDEBAR 1.3
Time Frames Used by the Committee in Its Deliberations

The report from the workshop held at the outset of this study (NRC 2007c) set forth the following time frames, which the committee used throughout its deliberations:

Short-term, 1- to 5-year R&D falls within the typical time span of a contract between EM and a cleanup contractor. This is also the typical time span for DOE and congressional funding plans and decisions. Short-term R&D is essential to solving problems that arise in the course of a cleanup activity, and it can lead to important and innovative solutions. It is usually funded by the contractors themselves.

Medium-term, 5- to 10-year R&D corresponds approximately to the time required to bring a promising result from applied research to technical maturity to provide a new approach to a cleanup problem. It could provide a safer, more efficient means of conducting an ongoing cleanup job, or a means to undertake a task for which a well-suited technology was previously unavailable.

Long-term R&D of greater than 10 years may be required to bring a completely new technology, perhaps resulting from an advance in science, to maturity or implementation. The practical applications of knowledge are hard to forecast. Nevertheless, exploratory research is the basic underpinning of truly new technologies—transistors rather than vacuum tubes.

As described in the committee's interim report, ensuring stable funding for medium- and long-term EM R&D is a necessary role of DOE headquarters and Congress. Given the 30-year time frame of the EM cleanup, the results of long-term R&D could be expected to provide large paybacks on investment by substantially improving EM's ability to conduct site cleanup.

Long term is also used in this report to refer to the time frames for which engineered solutions for waste containment are expected to remain effective. Typically these are time frames of several hundred years or longer. The degree of performance required of these containment systems and barriers for such long times is essentially unprecedented for engineered materials, such as concrete or grout. Understanding of the basic chemical and physical factors that govern such long-term performance is a good example of where medium- to long-term research is required.

Chapter 5 gives the committee's findings and recommendations. The chapter begins with a set of overarching considerations for EM's roadmapping that followed from the three key observations in the interim report, as noted earlier. The findings and recommendations are directed at future improvements of the Roadmap and how the Roadmap can help ensure continuity in EM's R&D programs, especially in maintaining staff expertise and infrastructure at the DOE sites and national laboratories visited by the committee. At the end of Chapter 5 the committee offers closing observations on how EM's Office of Engineering and Technology can enhance its role in leading EM's R&D program.

2

Principal Science and Technology Gaps

The first part of the statement of task for this study requests that the committee identify principal science and technology gaps and their priorities for the cleanup program. Previous National Research Council (NRC) reports have identified science and technology shortcomings using a variety of terms, for example, research needs, technology needs, cleanup challenges, and knowledge gaps (NRC 2007c). To address its task statement, the committee first sought an informative definition of the word "gap" (Figure 2.1).

The word "gap" is defined as a "discontinuity between two points" and in this context the task of identifying gaps could be interpreted to mean that the committee is to identify "showstoppers," that is, cleanup tasks for which there is insufficient knowledge or technology available to do the task. Information provided to the committee by the Office of Environmental Management (EM) and its contractors indicated that, if sufficient time and money were available to overcome cleanup obstacles, there are no showstopper gaps in the cleanup program. Another way of stating this is that EM and its contractors are confident that technologies EM has incorporated into its cleanup plans and schedules (baseline technologies) can be made to work.

Nevertheless, the committee observed in its interim report (Appendix H) that the complexity and magnitude of EM's cleanup task requires the results from a significant, ongoing R&D program if EM is to complete its cleanup mission safely, cost-effectively, and expeditiously. To address its statement of task in a way it judged would be most useful to EM, the committee chose as its working definition that *a gap is a shortfall in avail-*

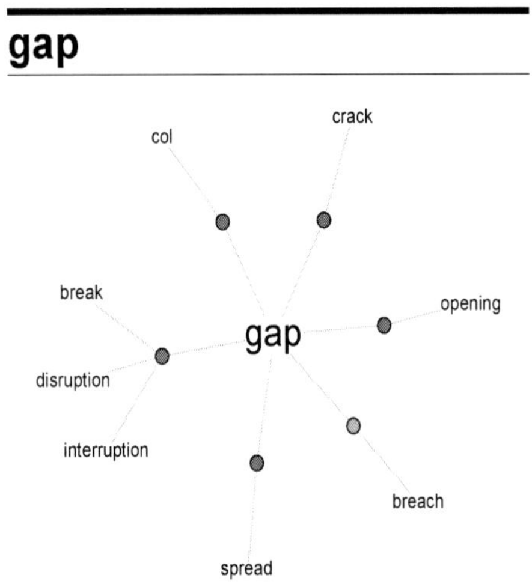

FIGURE 2.1 The word "gap" can be seen as the intersection of several synonyms. SOURCE: Visual Thesaurus: http://www.visualthesaurus.com/.

able knowledge or technology that could prevent EM from accomplishing a cleanup task on its expected schedule and/or budget. Following the analogy of a roadmap, a science and technology gap is a "pothole" in the road that EM might somehow work around, but at the likely cost of time and money. It would be much better to fill the pothole or avoid it altogether with appropriate research and development (R&D).

Addressing potholes could help EM to avoid large, insurmountable problems by addressing smaller technological challenges that could otherwise aggregate into showstoppers. Smaller investments in developing new science and technology could allow funding for several R&D approaches to a pothole problem, which would be more likely to lead to the most effective solution. Filling potholes before they erode into washouts is a natural role for EM's longer-term roadmapped research.

GAP IDENTIFICATION AND PRIORITIZATION

Gap identification began with the committee's March 2007 workshop and review of earlier Academies' reports (NRC 2007c), and then proceeded through the committee's site visits, which are summarized in the appendixes of this report. In this chapter, gaps are set forth as problems or potholes, which, if avoided or fixed with new technical tools, could help make the EM cleanup safer, faster, or less expensive. The identification of a gap does not imply that a given baseline technology might not work or should be abandoned—rather the gaps are incentives for EM to apply R&D to improve its available site cleanup and remediation tools. The titles of the gaps are intended to factually state a situation or condition that is less than optimal. The gaps have been identified at a level that the committee believes will provide EM the insights and flexibility to develop and implement effective, bounded, and targeted R&D to fill the gap.

EM's draft Engineering and Technology Roadmap, issued in April 2007, provided a framework for organizing the committee's fact-finding and deliberations,[1] but the committee worked independently of the specific contents on the Roadmap. Factors qualitatively considered when identifying the gaps included:

- Whether the gap required medium- to long-term R&D,[2]
- The volume of waste affected,
- Potential to reduce technical risks (including risk to workers),
- Reduction in schedule uncertainty,
- Potential cost savings,
- Likelihood of a successful outcome to the R&D effort, and
- Possible existence of solutions outside EM.

Applying these criteria to information received by the committee led to the general list of about 50 science and technology issues given in Appendix C. Later, through the course of its deliberations, the committee refined this list to the set of 13 principal gaps described in this chapter. In the committee's judgment, each of these 13 principal gaps could affect the schedule, cost, and risk associated with the EM cleanup program.

The priorities of the principal gaps were determined by the committee through an iterative process. During the August 2008 closed session, committee members who initially drafted gap analyses described the attributes of each gap to an "investment committee" composed of three committee members who previously held major programmatic and budgetary responsibilities.[3] The three suggested initial gap priorities according to the

[1] Essential features of the EM Science and Technology Roadmap are outlined in Chapter 1, Sidebar 1.2. For convenience it will be referred to as the EM roadmap or simply as the Roadmap throughout this report.

[2] These time frames are described in Sidebar 1.3.

[3] The three members of the investment committee were Carolyn Huntoon, former Department of Energy (DOE) Assistant Secretary for Environmental Management; Edwin Przybylowicz, former vice president for research at Eastman Kodak; and Andrew Sessler, former director of Lawrence Berkeley National Laboratory (LBNL).

information presented. The full committee then refined the gap analyses, prioritization criteria, and priorities. Gaps were prioritized as high, medium, or low within each program area (Table 2.1).

The committee did not attempt to prioritize the gaps across the program areas because the program areas differ fundamentally in the nature of

TABLE 2.1 Principal Science and Technology Gaps and Their R&D Priorities

Gap Number[a]	Statement of Gap	Priority
	Roadmap Program Area: Waste Processing	
WP-1	Substantial amounts of waste may be left in tanks/bins after their cleanout—especially in tanks with obstructions, compromised integrity, or associated piping.	High
WP-2	Low-activity streams from tank waste processing could contain substantial amounts of radionuclides.	Medium
WP-3	New facility designs, processes, and operations usually rely on pilot-scale testing with simulated rather than actual wastes.	Medium
WP-4	Increased vitrification capacity may be needed to meet schedule requirements of EM's high-level waste programs.	High
WP-5	The baseline tank waste vitrification process significantly increases the volume of high-level waste to be disposed.	Medium
WP-6	A variety of wastes and nuclear materials do not yet have a disposition path.	Low
	Roadmap Program Area: Groundwater and Soil Remediation	
GS-1	The behavior of contaminants in the subsurface is poorly understood.	High
GS-2	Site and contaminant source characteristics may limit the usefulness of EM's baseline subsurface remediation technologies.	Medium
GS-3	The long-term performance of trench caps, liners, and reactive barriers cannot be assessed with current knowledge.	Medium
GS-4	The long-term ability of cementitious materials to isolate wastes is not demonstrated.	High
	Roadmap Program Area: Facility Deactivation and Decommissioning	
DD-1	D&D work relies on manual labor for building characterization, equipment removal, and dismantlement.	High
DD-2	Personal protective equipment tends to be heavy and hot and limits movement of workers.	Low
DD-3	Removing contamination from building walls, other surfaces, and equipment can be slow and ineffective.	Medium

[a]Referred to throughout this report.

the risks that R&D could mitigate and in their timescales. Establishing their relative priorities involves policy judgments that are outside the committee's expertise. These differences and trade-offs are elaborated briefly below.

1. Cleanup work that involves only human activities, such as facility construction or demolition, can be accomplished on human-controlled schedules. Groundwater and soil remediation, on the other hand, involve geologic processes that humans can only attempt to control and, generally speaking, operate on a much longer timescale. Priorities for R&D, which typically include schedules and expected payoff, will be different for "engineering-only" projects versus those involving geologic processes (Sidebar 2.1).

2. The ultimate goal of site cleanup is to protect humans and the

SIDEBAR 2.1
Timescales for Engineering Projects Versus Geologic Processes

A conceptual incongruity exists in the time domains for DOE site cleanup between those activities associated with tank closure, waste separation and processing, and demolition and decommissioning and those activities associated with groundwater and soil contamination, environmental remediation, and long-term site stewardship. The pace of activities associated with the former is limited primarily by budget resources, which are manifested as physical infrastructure, size of the labor force, and availability of chemical and engineering solutions. Public and regulatory policies also modulate the selection and implementation schedules, such as tank closure. On the other hand, activities associated with soil and groundwater protection and cleanup are substantially controlled on a geologic scale by natural process rates and characteristics such as permeability of aquifers and the vadose zone, and the massive volumes of contaminated material (albeit at much lower concentration of problematic toxic organics, metals, and radionuclides than that of the original source materials). Simply put, removing waste from a tank is a straightforward, although complex, engineering challenge that can be addressed as such; soil and groundwater remediation cannot be similarly planned and accomplished.

A billion-dollar investment in tank closure or in a waste processing facility is likely to have a dramatic effect within a few years, while a like investment in groundwater remediation may only marginally accelerate the schedule for site cleanup and closure. The end state for the groundwater remediation at a site may be ultimately determined by an acknowledgment and acceptance that the site cannot be returned to a pristine state but that the residual contamination is sufficiently well understood scientifically that the risk to the public, end users, or the environment is acceptable, or can be reduced or controlled at acceptable levels far into the future.

environment from long-term consequences of the nation's former production of materials for nuclear weapons. Essentially this means remediating contaminated groundwaters and soils and protecting them from future contamination, which would argue that the higher priorities be given to the groundwater and soil gaps. However, the successful recovery and processing of waste from the former production operations, especially the high-level tank waste, is a prerequisite for preventing additional releases of contamination to the environment—in both the near and long terms. Facility decontamination and demolition, no matter how carefully it is planned and conducted, carries the potential for immediate injury or fatality among workers—as opposed to possible future consequence from groundwater and soil contamination. The program areas are thus intertwined and none rises to a higher priority than the others.

SCIENCE AND TECHNOLOGY GAP ANALYSES

The science and technology gaps that are presented in the remainder of this chapter are arranged according to the main program areas in the EM roadmap (DOE 2007a, 2008b). Each gap is analyzed in terms of how it is an obstacle for EM, and R&D opportunities to deal with it are described, as follow:

- An overview of the nature of the gap,
- The impact the gap has on EM's cleanup program,
- The current status of work related to the gap, and
- Future R&D approaches that EM could consider to help bridge the gap.

A table at the end of each gap analysis shows the basis for the committee's assessment of the gap's priority. Factors qualitatively assessed for each gap were volume of waste affected, potential to reduce technical uncertainty, potential to affect cleanup schedule, and potential to affect cost. To illustrate how these factors were evaluated, the millions of gallons of high-level tank waste ranked as high in the volume category, as did contaminated groundwater. R&D for gaps that reflected lack of knowledge tended to rate high for reducing technical uncertainty. Schedule and cost reductions are not always correlated, for example, for groundwater and soil remediation or waste treatment processes that are already under way.

WASTE PROCESSING

The waste processing program area of the EM roadmap deals primarily with high-level tank waste issues, including waste storage, waste retrieval, tank closure, waste pretreatment, and waste stabilization. Millions of gallons of high-level waste (HLW) from reprocessing nuclear fuels to recover plutonium and other nuclear materials arose during the Cold War era. Hanford, the first site that reprocessed fuels on an industrial scale, used several different reprocessing technologies that resulted in a variety of waste compositions. The Savannah River Site (SRS) mainly used one type of reprocessing technology and has a relatively smaller spectrum of waste compositions. Reprocessing activities at Idaho were at about one-tenth the scale of those at Hanford or SRS. There were important similarities and differences in waste management practices among these three sites, which are reflected in the science and technology gaps in waste processing identified by the committee and described in this section.

Waste Processing Gap 1 (WP-1): Substantial amounts of waste may be left in tanks/bins after their cleanout—especially in tanks with obstructions, compromised integrity, or associated piping.

Waste from former weapons material production at the Hanford site (Appendix D) is stored onsite in 149 single-shell (single-walled) and 28 double-shell tanks. The single-shell tanks were constructed between 1943 and 1964. The last of the double-shell tanks was constructed in 1986. All of the double-shell tanks have capacities of 1 million gallons. In total, 133 of the tanks have capacities of 500,000 to 1 million gallons. The Hanford tanks currently hold about 54 million gallons of waste, which contain a total of about 193 million curies of radioactivity (NRC 2006b). Hanford tank waste is very heterogeneous, but generally speaking it consists of supernatant liquid, water-soluble salt cake, and insoluble sludge. These phases resulted from the original acidic reprocessing waste being made alkaline for compatibility with the waste tanks, which were built from carbon steel, and subsequent evaporation of water to reduce the waste volume. The phases are layered and intermixed to varying degrees. Sludge removal is the most difficult.

SRS has 49 tanks in service that hold about 36 million gallons of waste containing about 426 million curies of radioactivity (Appendix G). The SRS tanks have a variety of designs—some single-shell, some double-shell, and some with the secondary shell less than the full height of the primary tank (i.e., "cup in a saucer"). The tanks vary in capacity from 750,000 to 1.3 million gallons. Most of the SRS tanks have internal cooling coils that were used to keep the temperature of the waste below boiling (NRC 2006b). SRS tank waste is broadly similar to that at Hanford although it is less heterogeneous chemically (Figure 2.2).

Most of the highly radioactive waste at the Idaho National Laboratory (INL) site is in the form of granular solids, which are stored in sets of stain-

FIGURE 2.2 Sludge sampled from an SRS tank. Tank sludge was formed by neutralizing acidic waste from the reprocessing of nuclear fuels and recovery of nuclear materials. This sludge flowed like a thick paste. Other sludges are more viscous or nearly solid, which makes them difficult to remove from the tanks. This approximately 2-liter sample was opened inside a shielded laboratory cell like that shown in Figure 3.2 in the late 1970s.
SOURCE: Department of Energy.

less steel bins contained in concrete vaults (Appendix E). The calcine waste exhibits a variety of sizes and compositions. It was originally transferred pneumatically into bins for storage, and DOE plans to retrieve the calcine essentially the same way. However, pneumatic retrieval could be difficult if the calcine has caked (e.g., from moisture in the bins or by particle-to-particle sintering). According to a presentation to the committee during its site visit, INL used a simulated calcine to demonstrate technical approaches for removing the binned calcine (Hagers 2007). The INL site still has about 900,000 gallons of acidic liquid waste stored in three stainless steel underground tanks.

According to the Ronald Reagan National Defense Authorization Act of 2005, Section 3116, for the tanks and their associated piping at SRS and Idaho to be closed, waste must be removed as much as is practical and meet the performance objectives in 10 CFR 61.40.[4] Closing a tank marks the end of EM cleanup activities for that tank.[5] Tank closure is EM's top priority for site cleanup, and it is a top priority among public citizens and their representatives.

The criteria under which Hanford tanks can be closed has not yet been established (NRC 2006b, Johnson 2008). Most of the legacy waste tanks at Oak Ridge were closed years ago although a few small surge and collection tanks remain (Appendix F).

Impact of the Gap

Tanks containing waste heels that have not been removed to the "extent practical" according to the Reagan Act or that cannot be shown to meet specified performance objectives to limit long-term radiation exposure cannot be closed. Tanks containing appreciable amounts of residual waste (heels) are unlikely to be accepted by DOE, its regulators, or the public for closure.

Removal of the bulk of the waste with large pumps (for SRS and Hanford) or pneumatic devices (for INL) appears to be relatively straightforward and efficient. However, experience at Hanford and SRS has shown that sludge heels inevitably remain in the tanks after the bulk of the waste has been retrieved. Reducing the volume of this heel becomes increasingly difficult, time-consuming, and expensive as the volume of the heel declines.

The tanks at Hanford and SRS generally have small access ports (risers); some tanks contain debris, and at SRS cooling coils further inhibit access and waste retrieval (Figure 2.3). A number of single-shell tanks at Hanford have leaked waste into the environment, and some double-shell tanks at SRS have leaked waste into the annulus between the tank walls (Figure 2.4). The structural integrity of tanks that have leaked is considered to be compromised. Buried waste transfer lines and ancillary equipment (e.g., smaller tanks, valves, transfer pits, and pumps) also contain waste.

---

[4] If these criteria are met, DOE can designate the residual waste as "waste incidental to reprocessing," a legal distinction that allows it to be permanently disposed onsite. Otherwise, classified as HLW, it would have to be removed for disposal in a licensed repository such as that proposed at Yucca Mountain, Nevada.

[5] Actions to close a tank after the waste removal criteria are met include isolating it from the waste system and filling it with a material such as grout with no intent for further waste retrieval.

FIGURE 2.3 Cooling coils in an SRS tank. Such coils maintained the temperature of high-level radioactive waste below boiling. The coils are an obstacle to removing the tank waste at SRS.
SOURCE: Department of Energy.

Current Status

Oak Ridge completed cleaning eight concrete-walled tanks in 2001, and all together closed 65 tanks between 1995 and 2007 (NRC 2007c). These tanks are smaller than those described above, but nonetheless demonstrated the use of several types of innovative remotely operated equipment, which led to substantial savings in cost and schedule (Boyd 2008). At the time of the committee's visit, Hanford had retrieved the waste from seven single-shell tanks, and waste retrievals were in progress or planned for four others (Mauss 2007). SRS has closed two tanks and is expected to have four more ready for closure by 2010. None of these tanks had internal cooling coils or other significant obstructions. The cleaning of a tank annulus has not been attempted.

Both Hanford and SRS operate tank mock-ups in which waste-retrieval challenges are simulated and new technologies are tested. In 2005 an EM subcontractor successfully retrieved simulated calcine from a bin (AEATES

FIGURE 2.4 Salt accumulated in a tank annulus. Double-walled tank construction helped to prevent the release of radioactive waste into the environment. In this figure minor leaks from the primary wall (right) have accumulated in the annulus. SRNL recently developed a robotic crawler for cleaning the tank wall.
SOURCE: Department of Energy.

2005). The sites have little experience in removing waste from bins, tank annuli, transfer pipes, or ancillary equipment.

Approaches to Bridge the Gap

Residual waste retrieval from tanks and ancillary pipelines was identified as an important technology gap in three NRC reports (2001b, 2003, 2006b). These reports recommended the development of physical and chemical cleaning technologies to improve the effectiveness of residual waste removal in tanks, tank annuli, and pipelines, especially technologies that reduce the risks of leakage of wastes to the environment during the removal operations (e.g., by using little or no water to retrieve wastes). Opportunities for expanding the use of robotics technologies for waste retrieval and tank cleaning are discussed in NRC (2006b). Site presentations at Hanford (Honeyman 2007) and SRS (Davis 2008; Spears 2008) included a number of technology needs for improving waste retrieval (Appendixes D and G).

According to this committee's assessment of information it received, the following approaches have promise for future EM R&D:

1. Chemical approaches that do not degrade the tanks or cause downstream problems, but that can dissolve recalcitrant (or agglomerated) solids in nonflowing areas (e.g., behind cooling coils, in clogged pipes). This could include R&D to mitigate the downstream effects of known chemical approaches. More extensive knowledge in the areas of (i) structure and dynamics of the materials and interfaces of relevance to the waste tank chemistry, (ii) complex solution-phase phenomena, and (iii) coupled chemical and physical processes might lead to transformational engineering solutions to this problem,
2. More autonomous physical approaches (e.g., focused water jets, grinders, pushers) to break up agglomerated waste and remove waste from surfaces while minimizing water use,
3. Faster and more autonomous physical approaches to corral solid materials (e.g., pushers) and to remove them from the tank while keeping water volumes low, and
4. Efficient approaches to demolish and remove internal tank structures to allow access for waste retrieval and reduce water intrusion pathways.

The sense of this committee, as well as the previous Academies' committees that considered EM's challenges with tank residues (NRC 2001b, 2006b), is that the sites need a variety of technologies—a toolbox—that can be applied on an as-needed basis to maximize EM's ability to retrieve its variety of waste types under a variety of tank conditions.

Prioritization of the Gap

Relative to other science and technology gaps discussed in this section the committee prioritized this gap as High.

	Relative Rating		
Criteria	High	Medium	Low
Volume of waste affected		X	
Potential to reduce technical uncertainty	X		
Potential to affect cleanup schedule	X		
Potential to affect cost	X		

WP-2: Low-activity streams from tank waste processing could contain substantial amounts of radionuclides.

The separation of the tank waste into high-activity and low-activity streams is a key to EM's plans for dealing with the waste (NRC 2006b). The sludge in Hanford and SRS tanks contains most of the waste radionuclides but comprises only about 10 percent of the tank waste volume at each site. The tank waste supernate and salt cake comprise about 90 percent of the waste volume (WP-1 and Appendixes D and G). Disposing of the salt cake and supernate (collectively called "salt waste"[6]) as HLW would increase HLW treatment and disposal costs—which are already the largest single component of the EM cleanup—about 10-fold.

The salt waste, however, contains most of the radiocesium (Cs-137) and small but significant amounts of Sr-90, Tc-99, and transuranic elements (TRU) such as plutonium. Accordingly, the supernate and dissolved salt cake will be processed before being stabilized ("pretreated") to remove key radionuclides from this soluble waste and route the nuclides into the HLW stream. At SRS the low-activity waste (LAW) stream is to be incorporated into a cementitious grout, referred to as saltstone, for permanent onsite disposal.

SRS encountered a significant obstacle in the mid-1990s when its planned salt processing technology was halted due to safety concerns. After several years of R&D, EM selected a new technology called "caustic-side solvent extraction" (CSSX) for SRS salt processing (Moyer et al. 2005). Development of CSSX began at Oak Ridge National Laboratory (ORNL), and it has continued with EM support at other national laboratories, especially the Savannah River National Laboratory (SRNL). The key new CSSX facility—the Salt Waste Processing Facility (SWPF)—is scheduled to begin operating in 2013 (Appendix G). Because its waste tanks are filled nearly to capacity, SRS has developed interim measures for processing its salt waste in order to free up tank space until the SWPF becomes available. One of these, the modular caustic-side solvent extraction unit (MCU), is essentially a pilot-scale test of CSSX with actual tank waste.

SRS operates its liquid waste facilities under State of South Carolina permits that impose a limit of 1.4 million curies in saltstone. Operation of the MCU and another interim process, referred to as deliquification, dissolution, and adjustment, will put about 1.2 million curies into the saltstone, while pretreating only a small fraction of the salt waste (Appendix G). Thus DOE has a very high expectation for the performance of the SWPF—to add only 0.2 million curies of radioactivity to the saltstone while processing

[6] Salt waste is primarily sodium nitrate, sodium nitrite, and sodium hydroxide.

the vast majority of the SRS salt waste.[7] This appears to be a significant technical risk.

The difficulties at SRS suggest that Hanford, with its more diverse tank waste compositions and characteristics, is likely to face challenges in processing its liquid wastes. Hanford has selected a newly developed ion exchange resin to remove radiocesium. While there has been extensive development work on this resin, the process has not been demonstrated in actual production.

Impact of the Gap

The SRS tank closure program cannot proceed without the ability to meet radionuclide separation objectives for its salt waste. Salt processing methods and objectives for the Hanford Waste Treatment Plant (WTP) are still in development. Baseline technologies are developed for both sites, but neither site has demonstrated the expected separations in actual plant operations. If the expected high separation factors are not achieved in actual operations, substantial program delays would be likely. There has been no decision as to whether Idaho calcine will require processing.

Work in Progress

In addition to developmental work on CSSX, SRNL has also made considerable progress in improving sorbents to be used to remove strontium and TRU from salt waste. These radionuclides contribute much less radioactivity to the salt waste than does cesium, but unless removed they could prevent the salt from meeting criteria that allow it to be disposed onsite. Along with the MCU, SRS is operating an actinide removal process (ARP) for its interim salt processing. The ARP will become a "front end" to the SWPF. Both Hanford and SRS are seeking higher-performance sorbents for salt processing (SRNL 2007; Tamosaitis 2007).

SRS is also considering a technology called small column ion exchange, which uses the ion exchange resin being developed at Hanford, to accelerate its salt processing. Ion exchange columns would be inserted into risers in a waste tank and fed from the tank itself—saving much of the cost of constructing a processing facility. Another recent technology development, the rotary microfilter, would be used to prepare a solids-free feed to the columns. This process was described as being near technical maturity (Davis 2008).

---

[7] It is not clear if radioactive decay is included in these numbers. Because of its 30-year half-life, about half of the Cs-137 in today's waste inventory will have decayed by the end of the SRS site cleanup program.

Approaches to Bridge the Gap

Baseline separation processes and materials may not be as efficient or tolerant of impurities or process upsets as expected. Based on this committee's assessment of information it received, the following have promise for future EM R&D toward making LAW processing more robust and efficient:

- Solvent extraction processes are susceptible to performance degradation from impurities. For example, silica, which is recycled in waste from borosilicate glass production in the Defense Waste Processing Facility (DWPF), can lead to poorly separable emulsions. Improved understanding of CSSX from operation of the MCU processes would be useful in increasing the efficiency of the MCU as well as ensuring that options are found for any operational issues that might arise. This, in turn, could help ensure that the SWPF meets its stringent performance requirements as discussed earlier.
- Organic ion exchange resins for cesium removal can be regenerated and reused, but are more readily degraded by radiation and corrosive chemicals than inorganic ion exchange material. Current inorganic ion exchange material generally cannot be regenerated and thus could contribute significantly to the waste volume. R&D to improve the lifetime of the organic resins or develop an elutable inorganic resin would address these problems and significantly enhance the efficiency of pretreatment.

Prioritization of the Gap

Relative to other science and technology gaps discussed in this section the committee prioritized this gap as Medium.

	Relative Rating		
Criteria	High	Medium	Low
Volume of waste affected	X		
Potential to reduce technical uncertainty		X	
Potential to affect cleanup schedule		X	
Potential to affect cost	X		

WP-3: New facility designs, processes, and operations usually rely on pilot-scale testing with simulated rather than actual wastes.

EM and its contractors are challenged with designing, building, and operating large, expensive, one-of-a-kind waste processing facilities that have major inherent safety risks because of the nature of the waste to be processed. The Hanford WTP and SRS SWPF are examples. In addition to being highly radioactive and chemically hazardous, the wastes exhibit a wide range of chemical and physical properties.

Developing waste treatment processes and scaling them up with assurance that they will work in production with actual waste feed streams is difficult. Working with actual waste in laboratory hot cells is generally on the gram to kilogram scale, which may not yield accurate data for process scale-up, and is cumbersome and slow. R&D at the pilot scale using actual radioactive materials can be exceedingly expensive—requiring engineering and construction work similar to building the full-scale facility itself—and therefore usually not feasible.

To avoid these pitfalls R&D normally proceeds along two complementary paths: bench-scale hot cell work with radioactive materials and pilot-scale work using nonradioactive simulants. However, developing a simulant that accurately represents the characteristics of each radioactive waste composition can take significant effort, especially for sludges. Because the composition of the sludge in any given tank may vary significantly, and different simulants may be required for different aspects of a single sludge (e.g., rheological properties, chemical properties, radiolytic properties), multiple simulants can be required for a particular waste stream.

The committee heard of numerous basic waste processing operations that carry significant technical risk because they cannot be tested on a pilot-plant scale with actual wastes, including:

- Reliable separation of solids from liquid waste streams to prevent clogging of ion exchange beds or adverse effects on solvent extraction equipment,
- Ensuring that shear-thickening (non-Newtonian) sludges can be transported in pipelines without clogging,
- Predicting the rate of radiolytic hydrogen generation by process sludges and the release time and rate of the hydrogen,
- Predicting the stability and interaction of various process streams to allow for reduced conservatism in the operational safety bases of the tank farms, and
- Understanding the effect of impurities and degradation or corrosion products on process performance.

Impact of the Gap

Failure to accurately predict the flow properties of slurries is a leading cause of process failure (Merrow 2000). The absence of adequate understanding of the behavior of process streams can necessitate overly conservative and costly process designs to minimize the risk of a process failure or the risk of unrecognized safety issues, which as a worst case can render a facility inoperable with the actual radioactive waste it was intended to process.[8]

An overly conservative process flowsheet can prevent efficient operation of tank farms (e.g., prevent some wastes from being combined or processed), which increases costs and the time required for tank cleanup. The use of trial-and-error methods to develop representative simulants is costly and tends to increase reliance on even more costly hot cell operations. Additionally, many simulants contain hazardous materials that lead to high disposal costs for the simulants and equipment contaminated with them.

Work in Progress

Pacific Northwest National Laboratory (PNNL) is building a quarter-scale engineering "platform" to test and demonstrate WTP pretreatment operations using simulants—including sludge washing, leaching, and waste concentration (Figure 2.5). PNNL also has an extensive program to test pulse jet mixers, which are key components for mixing solid/liquid slurries in several WTP operations, with simulants. PNNL considers the liquid-solid mixing problems in design of the WTP to be on the forefront of mixing science (Michener 2007). EM has sponsored hot cell R&D to improve understanding of the in-process behavior of key radioactive materials at all four of the national laboratories visited by the committee.

EM, in conjunction with the national laboratories, has begun to use lab- and pilot-scale data to verify and calibrate computational fluid dynamic (CFD) or other types of numerical models. These verified models can then be exercised through the range of conditions that might be encountered by varying parameters (e.g., flow rate, viscosity, solids loading, gas loading) to see what the effect on equipment operation is. Based on this information, the process design can be modified as necessary. The use of computer modeling verified through engineering tests is the mainstay of most industrial organizations and especially industrial chemical producers.

Approaches to Bridge the Gap

According to this committee's assessment of information it received, computer modeling (e.g., CFD) can help bridge the gap between data that can be obtained from lab-scale tests with actual wastes and pilot-scale tests

---

[8] A worst case happens rarely. However, one example is the SRS in-tank precipitation process for radiocesium removal, which behaved unexpectedly during a full-scale test with actual radioactive waste and was abandoned (NRC 2000a, 2001a, and Appendix G).

FIGURE 2.5 Diagram of pretreatment semiworks for the Hanford WTP. Such pilot-scale semiworks are essential for testing new processes before they are used to treat actual radioactive waste in large, expensive new facilities. Semiworks testing is done almost exclusively with nonradioactive simulants. The operators depicted on the right indicate the size of this semiworks.
SOURCE: Department of Energy.

with simulants. However, the use of computer modeling to replace large pilot- and full-scale testing with simulants carries some technical risk.

These technical risks could be reduced if CFD or other models of relatively complex behaviors could be calibrated using data from tests with actual wastes. The models would then be used to predict the fluid system's behavior under other conditions. Engineering tests under those conditions would determine the degree to which the computer-generated predictions were met. This approach could be used for a number of different phenomena including heat transfer, fluid flow in tanks and porous media, explosive atmosphere testing, chemisorption phenomena on resins and other solid media, and precipitate formation in heat exchangers and on pipes, pumps, and vessels.

An essential component of bridging the gaps among waste simulants, computer models, and the behavior of actual waste will be R&D aimed at discovering potential, unexpected interactions or other phenomena inherent in the actual wastes that could lead to a process upset or failure. Such discovery-oriented R&D would help ensure that the conceptual model, which is manifested by the computer model, is correct.

Prioritization of the Gap

Relative to other science and technology gaps discussed in this section the committee prioritized this gap as Medium.

Criteria	Relative Rating High	Medium	Low
Volume of waste affected		X	
Potential to reduce technical uncertainty		X	
Potential to affect cleanup schedule			X[a]
Potential to affect cost	X		

[a]Addressing this gap would significantly impact only new construction in the complex. It could also provide technical support or lead to modification of facilities that are finished or well along in construction and design, such as the SWPF and WTP.

WP-4: Increased vitrification capacity may be needed to meet schedule requirements of EM's high-level waste programs.

A Joule-heated melter is being used at SRS to stabilize high-activity tank waste in borosilicate glass. Hanford is taking a similar approach, but the Hanford WTP is being designed to vitrify both high-activity and low-activity streams from its tanks using different melters for each stream. Joule-heated melters might be used at Idaho depending on future decisions concerning the disposition of calcine.

The design and operation of Joule-heated melters for vitrifying tank waste limits their throughput. A slurry of waste and water is added to a "cold cap" on top of the molten glass. Water evaporates from the slurry, forming more of the cold cap. Material at the bottom of the cold cap is gradually incorporated into the melt. Heat to produce the glass comes from passing electricity directly through the melt (i.e., the Joule effect). Mixing within the melter is convection-driven and inherently slow in the viscous melt.

New approaches such as bubblers are planned for use at the WTP to aid in mixing the melt and increase throughput. Although bubblers can increase the rates by up to 50 percent, other operational issues arise such as increases in the amount of volatilized chemicals and radionuclides that would have to be trapped in the off-gas treatment system. Even with bubblers, the WTP would have the capacity to vitrify only about one-third of Hanford's LAW. This is driving Hanford's plan to use a supplemental treatment process for the additional LAW. If additional capacity can be obtained in the current space allocated for the WTP melters, then the need for supplemental facilities could potentially be avoided.

Impact of the Gap

The throughput of a Joule-heated melter is relatively low and increasing the size is not practical since the physical installations are already complete. The number of years that the WTP and DWPF will operate depends on the time it takes to vitrify the waste on the respective sites. With each additional year of operation costing about $500 million in today's money (Davis 2008), increasing the melter throughput by a factor of 2 could potentially save several billion dollars on each site.

Tamosaitis (2007) described several factors, in addition to the large volume of waste, that make WTP throughput a concern. These include the diversity of waste input streams, behavior of solids in the system, and process upsets. He listed improved waste forms, glass formulations, and melters as technology needs for enhancing throughput.

At SRS both Spears (2008) and Davis (2008) listed increasing the throughput of the DWPF as technology needs. Davis stated that options needing further R&D include improving the glass-forming frit, improving the ability to mix the contents of the melter, and operating at a higher temperature with an alternative melter design.

Work in Progress

Hot-wall induction melters are in use in France and the United Kingdom, and R&D on cold-wall induction melters is being performed in France and Russia (Ahearne 2002). INL researchers showed the committee a high-throughput induction melter that is being used for R&D with simulated wastes (Appendix E). Similar work has been performed at SRS based on operations at France's Marcoule site (Barnes et al. 2008). Plasma-based melters that were capable of high throughputs within a small footprint were also examined by Westinghouse (McLaughlin et al. 1994, 1995). In addition, other R&D such as the bubbler work at the Catholic University of America and Russian work using microwaves to help heat the cold cap of the melter (Kurkumeli et al. 1992) have been carried out. Considerable R&D on alternative stabilization technologies for Hanford LAW, including bulk vitrification and steam reforming, is also ongoing at INL and Hanford. NRC (1999a) recommended that DOE examine a range of technical options for immobilizing HLW calcine at Idaho if the calcine itself is not adequate to meet final disposal requirements.

Future Approaches to Bridge the Gap

SRS has a current need to increase the capacity of the existing DWPF melter to meet current programmatic needs. Simply installing a bigger melter is precluded by the design of the DWPF. At Hanford more capacity is needed, especially for the large volume of LAW. The WTP at Hanford is in the advanced design stage.

Based on this committee's assessment of the information it received, the following have promise for future EM R&D:

- Alternative melter designs, with special attention to induction melters or other types of melters that have high throughput relative to their size.
- New methods of enhancing mixing in Joule-heated melters to increase their capacity. Bubblers can improve mixing, but they have potential technical problems associated with corrosion/erosion of the bubbler tubes and other components, and disturbing the cold cap.[9]
- Alternatives for boosting heat input to Joule-heated melters. Microwave heating might be one option (NRC 2005).
- New glass frit formulations that have lower viscosities to allow improved convective heat transfer in Joule melters, for example, adding Li_2O to the frit. Any new formulation will have to be tolerant of variations in the waste stream.

Prioritization of the Gap

Relative to other science and technology gaps discussed in this section the committee prioritized this gap as High.

	Relative Rating		
Criteria	High	Medium	Low
Volume of waste affected	X		
Potential to reduce technical uncertainty		X	
Potential to affect cleanup schedule	X		
Potential to affect cost	X		

[9] Disturbing the cold cap would increase the load of volatile materials and radionuclides on the melter off-gas system or possibly trap water beneath portions of the melt, which could lead to eruptions of steam and molten glass.

WP-5: The baseline tank waste vitrification process significantly increases the volume of high-level waste to be disposed.

Along with the rate at which a Joule-heated melter can produce the borosilicate glass waste form discussed in WP-4, the percent of waste that can be incorporated into the glass determines the rate at which tank waste can be processed (vitrified) for disposal. There are a variety of factors that determine the amount of waste that can be incorporated into a given volume of glass (waste loading). These factors include the temperature of the melt, the composition of the glass-forming material (frit), and the composition of the waste.

Tank waste sludge contains some constituents, such as sulfate (SO_4^{2-}), phosphate (PO_4^{3-}), and chromium, that are relatively insoluble in the glass melt. The waste also contains aluminum, which increases the viscosity of the melt, making it hard to pour from the melter, and sodium, which reduces the durability of the glass. Generally speaking, the best sludge loadings that are currently expected under production conditions with currently available vitrification flowsheets are in the range of about 30 to 40 weight percent dry sludge with 60 to 70 weight percent of the added frit. In other words, current HLW forms are made from about one-third waste and two-thirds binder.

Impact of the Gap

Coupled with the limited throughput of current melters, the low waste-to-glass ratio establishes a decades-long time frame for working off the high-activity waste inventories at Hanford and SRS. Tank operations and a good deal of site infrastructure will have to remain open and operating to support waste vitrification during these decades. The approximately $500 million per year cost at each site for maintaining its high-activity waste operations is a strong incentive for faster waste processing (Appendixes D and G). A prolonged waste processing schedule also increases the chance that some tanks storing the waste, which have already exceeded their design lives, will leak.

Current Status

There are two approaches for increasing glass waste loading that are being investigated at Hanford, PNNL, SRS, and SRNL. One is to modify the glass formulation so that more waste can be incorporated into the glass matrix without compromising its processability or quality of the product. EM-supported work on improved frit formulations has allowed the waste loading of glass produced by the SRS DWPF to be increased from roughly 30 percent to around 38 percent.

The second approach is to pretreat the waste to remove bulk nonradioactive constituents (aluminum, iron, sodium) to reduce the waste volume to be vitrified or to remove waste constituents that have low solubility in borosilicate glass (chromium, sulfate) and thus limit waste loading. R&D concerning selective removal of nonradioactive, solubility-limiting constituents from the sludge is being pursued at SRS and Hanford. Removal approaches have focused on water washing to remove sodium salts and washing with caustic solutions to remove aluminum and chromium.

Future Approaches to Bridge the Gap

According to this committee's assessment of information it received, the following have promise for future EM R&D:

- Additional work on understanding the chemical nature of nonradioactive components in high-activity waste that might lead to their selective removal. Some forms of aluminum can be removed readily by caustic washing of the sludge, which has been demonstrated at SRS. Removal of more recalcitrant forms of aluminum may be necessary for aluminum removal to become a practical way of reducing waste volume, especially at Hanford where tank conditions (heat, age) have probably produced more of the recalcitrant forms. Chromium in Hanford waste will reduce waste loadings in WTP glass unless it is removed in pretreatment or reduced in concentration by blending with low-chromium waste.
- Work to develop entirely new, nonborosilicate glass waste forms that can accommodate higher waste loadings and/or loadings of problematic constituents like aluminum, chromium, and sulfate. Phosphate glasses were one alternative class of waste forms described to the committee.
- Waste forms that include little or no added binder. Idaho calcine is one such example. Perhaps sintered or minimally bonded sludges could be developed for Hanford and SRS. Such work would probably rely heavily on computer modeling of waste and repository characteristics to show that they could meet their disposal requirements. Ensuring that the Idaho calcine can be disposed without further treatment would provide a strong cost driver for this R&D. Processing Idaho calcine to produce a different waste form would likely require a DOE investment of several billion dollars.
- Work that is synergistic with WP-4. New melter technology might, for example, allow waste processing at higher temperatures, which could increase waste loading and provide faster throughput as well.

Prioritization of the Gap

Relative to other science and technology gaps discussed in this section the committee prioritized this gap as Medium.

	Relative Rating		
Criteria	High	Medium	Low
Volume of waste affected		X	
Potential to reduce technical uncertainty			X
Potential to affect cleanup schedule	X		
Potential to affect cost	X		

WP-6: A variety of wastes and nuclear materials do not yet have a disposition path.

Wastes and nuclear materials that do not have a defined disposition path are "orphans." Examples of orphans include:

- Over 4,000 cubic meters of calcine at Idaho that may require processing such as vitrification if it cannot be shown to be acceptable for in situ disposal or geologic disposal in its present form;
- Idaho's sodium-bearing tank waste, which is presently classified as TRU waste;[10]
- The waste left in SRS Tank 48 that contains tetraphenyl borate (used in a previous attempt to remove Cs-137 from tank waste) and its degradation products, which may require special processing to convert them into a stream suitable for vitrification at the DWPF;
- Spent fuel at Idaho for which adequate characterization to qualify it for disposal is impractical because of the high radiation field and lack of access to the nuclear materials in the sealed packages;
- Aluminum-clad N-reactor fuels at Hanford that may not meet criteria for disposal in a deep geologic repository because of their susceptibility to corrosion;
- K-basin sludge at Hanford that contains pyrophoric uranium metal;
- High-atomic-weight ("heavy") actinide targets at ORNL that have no further use;
- The used beryllium neutron reflectors at the High Flux Irradiation Reactor at ORNL and Advanced Test Reactor at INL, which are classified as civilian rather than as defense waste. Their content of TRU radionuclides from uranium impurities and/or carbon-14 from nitrogen impurities puts

[10] Idaho currently has a project for treating these wastes, but they will have to be stored onsite until their final disposition is decided.

them in the civilian greater-than-Class C (GTCC) waste category. Wastes in this category presently have no disposition path (NRC 2006a).

These wastes and nuclear materials have significant radioactivity and therefore hazards associated with them. Some of the wastes constitute a substantial volume or number of items that must be dispositioned. There may be enough variability in composition or other properties among wastes in any of the categories above to make their characterization a challenge.

Impact of Gap

The existence of this gap has two ongoing and potential impacts:

- Continuing cost and occupational doses result from having to continue to operate the storage facilities and associated site infrastructure.
- Until stabilized, many of these materials will continue to degrade, especially those stored underwater, which could increase the cost and hazard of retrieval, treatment, and disposition.

Current Status

Alternatives for disposition of the INL calcine and Tank 48 waste are being evaluated but the unique nature of these wastes and limited previous R&D suggest that a fundamental understanding of these materials is not in hand. R&D is under way to prevent future beryllium reflectors from becoming GTCC waste.

Approaches to Bridge the Gap

The NRC has previously recommended R&D (NRC 1999a, 2001b, 2003, 2005, 2006b), but little has been done to determine the final disposal route for these wastes. According to this committee's assessment of information it received, the following have promise for future EM R&D:

- A systematic effort to develop the technical basis for alternative characterization, treatment, and disposal options, and for waste acceptance criteria;
- A systematic effort to understand the degradation rate of nuclear materials in storage with initial focus on materials stored underwater;
- Risk-informed comparison of the alternatives for disposition of INL calcine and SRS Tank 48 waste; and
- Improved methods for characterizing highly radioactive spent fuel and nuclear materials inside containers.

Prioritization of the Gap

Relative to other science and technology gaps discussed in this section the committee prioritized this gap as Low.

Criteria	Relative Rating		
	High	Medium	Low
Volume of waste affected	X[a]		
Potential to reduce technical uncertainty		X	
Potential to affect cleanup schedule			X
Potential to affect cost			X

[a]If Idaho calcine in its present form is determined to be unacceptable for disposal.

GROUNDWATER AND SOIL REMEDIATION

The groundwater and soil remediation program area of the EM roadmap deals primarily with environmental sampling and contaminant characterization, treatment, remediation, and modeling to guide cleanup. EM is responsible for about 6.4 billion cubic meters of contaminated soil, groundwater, and other media that may require remedial action (NRC 2000c, p. 24).

Chemicals, metals, and radionuclides were introduced into the environment at DOE sites through accidental spills and leaks from storage tanks and waste transfer lines and also through intentional disposal via injection wells, disposal pits, and settling ponds. Releases into the environment generally were not closely tracked, and many release sites were unmarked and forgotten. Some of these sites are being rediscovered as EM proceeds with its cleanup program.

Chlorinated hydrocarbons or volatile chlorinated organic compounds are the most prevalent group of contaminants at DOE sites, appearing with a frequency of 82 percent in plumes reported in the EM Ground Water Database (GWD).[11] These compounds were used in large quantities as cleaning agents, solvents, or lubricants. Some of these compounds, which are sparingly miscible with water and denser than water—notably carbon tetrachloride—comprise a category of contaminants referred to as dense, nonaqueous phase liquids (DNAPLs). Organic contaminants are often co-contaminated with tritium or nitrates (Hazen et al. 2008).

Plutonium, uranium, strontium, technetium, chromium, and mercury are among the problematic radionuclides and toxic metals that have been difficult to predict in occurrence and transport. While metals and radionu-

[11] See http://www.em.doe.gov/Pages/groundwatersoildatabase.aspx?PAGEID=DB.

clides (other than tritium) are reported to occur in only about 5 percent of the plumes in the GWD, they occur together nearly 30 percent of the time and are associated with volatile organic chemicals (VOCs) approximately 25 percent of the time (Hazen et al. 2008). Different contaminants combined into clusters of three or four are not uncommon, and they frequently include radioisotopes, metals, sulfates, and nitrates. Such combinations probably reflect their origins from onsite chemical operations and subsequent interactions within the geologic media.

In 2004, a DOE Inspector General's audit (DOE 2004c) found the continued use of pump-and-treat technology to be relatively ineffective, that innovative groundwater contaminant monitoring is not being exploited, and that implementing current treatment and barrier technology may need abeyance until realistic end states are more sharply defined. Thompson (2007) reported that a 2006 audit by the Government Accountability Office found fault with DOE's remediation efforts to prevent contaminants from reaching the Columbia River. The audit (GAO 2006) concluded that technology used in several remedies is not performing satisfactorily, and that there is a lack of new technologies to address contamination issues. Groundwater remediation challenges are recognized in previous NRC (1993, 2000b, 2004) reports and were reiterated in the committee's workshop summary (NRC 2007c).

GS-1: The behavior of contaminants in the subsurface is poorly understood.

Geochemical and biochemical oxidation and reduction of metals and radionuclides can dramatically alter their solubility in groundwater, sorption on solid substrates, and colloidal transport properties. These are complex, dynamic, and often reversible processes that cannot be predicted without knowledge of the contaminant chemistry, the subsurface biogeochemical and hydrogeologic properties and their spatial variability, and dynamics of water recharge and removal (e.g., precipitation, stream flow, groundwater pumping).

Lack of basic understanding of contaminant and site characteristics can lead to incorrect concepts of contaminant behavior that have, in turn, led to a disconnect between the expected and actual outcome of remediation efforts. Several examples illustrate the importance of understanding basic contaminant biogeochemistry and characterizing the properties of the field site adequately when planning whether or how to conduct soil and groundwater remediation:

- At Oak Ridge, the levels of mercury in East Fork Poplar Creek, which is downstream from Y-12, have been reduced to meet drinking water standards by cleanup actions already completed. However, the mercury concentration in fish and aquatic life is continuing to increase (Appendix F), suggesting a critical aspect of the contaminant distribution, transport, and/or biotransformation at the field site is not known.

- Initial predictions of contaminant migration to the water table at Idaho's RWMC developed in the 1960s were on the order of 100,000 years. Improved knowledge of subsurface transport processes has led to travel time estimates that are on the order of decades (NRC 2000c, p. 30).
- Stewart (2007) reported seven examples of apparently anomalous contaminant migration at Hanford—the contamination was moving in unexpected amounts and/or unexpected directions. The reasons underlying the apparently anomalous behavior were resolved in each case by scientific study that led to improved approaches for remediation or containment of the contamination. However, she also noted problems for which scientific understanding is limited. One example is the deeper migration of plutonium into the vadose zone beneath the Z cribs than has been predicted with current site models.
- In the Hanford 300 area, uranium-contaminated soil was removed, and the plume was expected to meet the water quality standard within 10 years of the remediation. This did not happen. Incomplete characterization of the source zone and the consequent lack of a remedy to deal appropriately with the source zone have prevented EM from meeting the original cleanup time line. In addition, the assumption that the groundwater plume would dissipate during the 10 years following soil removal delayed further progress to understand the source of what is now known to be an ongoing groundwater contamination issue.
- The unexpected detection of chromium in a monitoring well at Los Alamos substantially reinforced concerns about the adequacy of that site's groundwater protection program (NRC 2007b).

Impact of the Gap

If the rate of progress or result of a remedial action is less than expected, it can delay schedules (including missed regulatory milestones), increase costs, and undermine stakeholder confidence in EM's site cleanup. Remediation carried out without complete (or ongoing) characterization of the contamination source and factors that control contaminant movements is a technical risk—simply removing some of the contaminant mass may not cause the expected response in contaminant concentration or movement. Without knowledge of the fundamental processes that interact to determine contaminant mobility and persistence in the spatially heterogeneous geologic settings that exist at each of the DOE sites, it is impossible to complete a reliable risk assessment or plan an effective remedial program.

Current Status

There is substantial R&D under way at the DOE sites and by many other organizations to better understand the science needed to predict how various hazardous substances are mobilized, transported through the geosphere and biosphere, and affect humans and the environment. For example, more detailed characterization of the source of the uranium plume in the Hanford 300 area is included in one of three Integrated Field-Scale Subsurface Research Challenges funded through the Environmental Remediation Science Program (ERSP) within the Basic Energy Sciences Office of DOE's Office of Science (SC).[12]

Work to improve conceptual and computational models of contaminant migration is ongoing within the Subsurface Science Focus Area through support by the Environmental Remediation Sciences Division of the DOE Office of Science's Office of Biological and Environmental Research (OBER).[13] Examples of Subsurface Science Focus Area projects include PNNL's research on the role of microenvironments and transition zones in the reactive transport of technetium (Tc), uranium, and plutonium;[14] ORNL's research on the biogeochemical transformations that govern mercury speciation at the sediment–water interface;[15] and LBNL's research to develop a sustainable systems approach for addressing critical knowledge gaps associated with environmental stewardship of metals and radionuclides in the subsurface.[16]

Organizations such as the DOE offices of Science and Civilian Radioactive Waste Management, the Environmental Protection Agency (EPA), the Nuclear Regulatory Commission (USNRC), and the U.S. Geological Survey all have technical expertise relevant to assessments and could be helpful in establishing generic approaches and assumptions. The EPA and USNRC both do technical assessments to support decisions that include substantial stakeholder involvement, and their experience could be useful to the DOE cleanup program.

The EPA Superfund program established to support cleanup and risk reduction of those sites on the National Priorities List (NPL) and the Superfund Innovative Technology Evaluation (SITE) program are helpful resources in addressing what works and what does not for problematic sites and in deployment of new technology.

While some contaminants and discharge magnitudes are unusual or unique to DOE sites, many industries and universities have major R&D programs for understanding contaminant fate and transport.

---

[12] See http://www.hanfordifc.pnl.gov.

[13] See http://ersdprojects.science.doe.gov/.

[14] See http://www.pnl.gov/biology/sfa/.

[15] See http://www.esd.ornl.gov/programs/rsfa/index.shtml.

[16] See http://esd.lbl.gov/research/projects/sustainable_systems/.

Approaches to Bridge the Gap

According to the committee's assessment of information it received, the following have promise for future EM R&D:

- Development of improved technology and methodologies for source and plume characterization and monitoring with emphasis on real-time analytical monitoring instruments for field use, remote data acquisition, and automated data collection;
- Continued study of contaminated materials from vadose zone and groundwater plumes to improve conceptual models of geochemical and biogeochemical processes (and species) controlling the mobilization and immobilization of contaminants, including complex chemical mixtures;
- Use of more sophisticated computational models that better incorporate understanding of site geohydrology and contaminant geochemistry; and
- Development of scientific bases to support delaying remediation activities until there is an adequate knowledge base to proceed with the remediation.

The technical challenges in groundwater and soil remediation differ from those in waste processing in terms of the timescales during which the relevant processes operate, access and ability to measure process parameters (reading a gauge versus ascertaining what is going on belowground), and ability to control the process parameters. The need for adequate characterization of site hydrogeology and contamination sources and plumes is recognized by EM and SC. Partnering with SC (Chapter 4) can provide EM with access to state-of-the-art science capabilities to improve its site characterization through more robust modeling and advanced instrumentation.

In the case of the EPA Superfund program, each NPL site has a Record of Decision (ROD) that describes how the given site will be cleaned up or managed. The RODs are readily accessible public documents that may translate to EM site closure. Perhaps more informative are the Remedial Investigation/Feasibility Study reports that lay out the investigative assessment of a site and the treatability studies required to select and support site cleanup.

The congressional language that led to the EM roadmap cited a previous NRC (2005) report, which found that "monitoring systems at EM closure sites have been estimated to be some 25 years behind the state-of-art." and that "an improved capability for environmental monitoring would strengthen EM's plans to leave waste and contaminated media at DOE sites." Two NRC reports (2005, 2007b) suggested R&D in noninvasive geophysical sensor techniques such as electromagnetic and electrical resistivity methods, seismic reflectivity, and ground-penetrating radar to reduce, as much as possible, the practice of physically sampling and analyzing groundwater samples, which is currently prevalent at DOE sites.

It has been argued that the complexity of the sites, microlevel of resolution needed, and integration among models require computational capacity that may not be justifiable or that model complexity is too great to provide much real value to practitioners. While this may have been true in the past, advances in both the computational capability within DOE and risk assessment modeling clearly indicate that new modeling efforts in mixed contaminant fate and transport analysis, in the development of methods for scaling up from micromeasurements to field-scale prediction, and in the simulation of remediation processes are promising. This computational power could be linked to the development of improved methods for characterizing site conditions, formulating conceptual models that represent system behavior, parameterization and calibration of site-specific models, and quantification of uncertainties in prediction. Modern computing power can help ensure that more sophisticated numerical models are well integrated with the biochemical, ecological, and geochemical sciences sufficiently to provide the resolution needed to improve the accuracy of model simulations and predictions needed to advance cleanup, remediation, and risk reduction. For example, promising work at the Environmental Molecular Sciences Laboratory at PNNL in the use of instruments such as nuclear magnetic resonance to generate microscale data on intragrain diffusion rates of dissolved uranium in tank-waste-contaminated sediment particles suggests that new modeling efforts can be appropriately parameterized.

There are some instances of which today's understanding of site and contaminant characteristics and/or available technologies are probably not adequate for a successful remediation program. The BC cribs on the Central Plateau at Hanford are likley examples. The liquid waste disposed at the BC cribs and trenches represents some of the most concentrated radioactive and hazardous waste disposed to the ground at Hanford. Based on inventory estimates, this site contains the largest inventory of technetium-99 in the Hanford soil.[17] The majority of the Tc-99 is believed to be located in the site's vadose zone, which comprises highly stratified glacial-fluvial sediments that give rise to complex subsurface-flow paths (Gee et al. 2007). This complexity, combined with uncertainty about the in-ground chemical and/or biological processes that influence Tc-99 behavior, means that the crucial information needed to design appropriate remedial actions for the site is missing.

Another example is the chlorinated organic contamination at the East Tennessee Technology Park at Oak Ridge (Appendix F). This case involves

[17] See http://www.hanford.gov/cp/gpp/functionalareas/wastesite/bccribs.cfm.

the presence of a DNAPL in a fractured bedrock aquifer, which is a combination of contaminant and subsurface characteristics that is universally acknowledged to be extremely challenging to effectively remediate. Oak Ridge listed a need for scientific and technical support for a Technical Impracticability (TI) waiver for remediating these source areas (Phillips 2007). A decision to grant a TI waiver represents regulators' concurrence with a finding that restoration of contaminated soil and/or groundwater to agreed cleanup levels cannot be achieved using currently available or new and innovative methods or technologies.[18]

For such cases in which current knowledge and technology are likely insufficient to ensure successful remediation, an alternative approach is the "cocooning" concept, which is being used for the Hanford reactors (NRC 2005). The concept of cocooning for soil and groundwater contamination problems that cannot be technologically addressed at present is to adaptively manage the contamination in order to avoid actions that involve costly or inappropriate treatment activities and result in little to no risk reduction.

Developing the science and technology base to show that a temporizing measure, such as pumping strategies to achieve hydraulic containment that prevents the spread of a plume (GS-2) or placing a water barrier over a trench (GS-3), is protective while continuing to pursue better solutions as the state of the art advances could help EM deal with currently intractable situations. R&D to demonstrate that the current situation is safe, for now, and to develop a temporizing remedy could save money and time and avoid the perception of failure. Roadmapping the longer-term research to eventually address the problem could help assure stakeholders that the problem is not simply being pushed aside.

Prioritization of the Gap

Relative to other science and technology gaps discussed in this section the committee judged the priority of addressing this gap as High.

	Relative Rating		
Criteria	High	Medium	Low
Volume of waste affected	X		
Potential to reduce technical uncertainty	X		
Potential to affect cleanup schedule	X		
Potential to affect cost		X	

[18] See: http://homer.ornl.gov/oepa/guidance/cercla/techimpract.pdf.

GS-2 Site and contaminant source characteristics may limit the usefulness of EM's baseline subsurface remediation technologies.

There are a wide variety of methods for remediating contaminated groundwaters and soils that have been developed and used by private industries and government agencies. These include technologies such as pump and treat, biostimulation and bioaugmentation, air-sparging, soil vapor extraction, electrokinetics, phytoremediation, in situ flushing and in situ oxidation, permeable reactive barriers (PRBs), in situ thermal treatment, multiple-phase extraction, and monitored natural attenuation (MNA). Among these, pump and treat is the most commonly used, and it appears to be favored by EM site cleanup contractors (Figure 2.6). It is a mature technology, frequently used for remediating groundwater contaminated with a variety of substances, including VOCs, residues of explosives, and dissolved metals. Contaminated groundwater is removed from the subsurface by pumping, treated to remove the contaminants, and returned to the aquifer or discharged.

The water well design, pumping system, and treatment depend on the site characteristics and contaminant type. Aboveground treatment technologies for extracted contaminated groundwater typically include biodegradation, filtration, air stripping, and adsorption. It is not uncommon to find many wells extracting groundwater at the same time. These wells may extract water from different depths to maximize effectiveness. Groundwa-

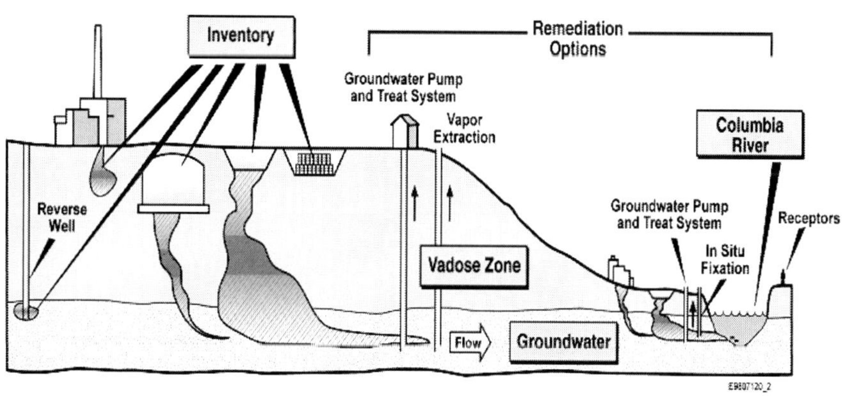

FIGURE 2.6 Illustration of groundwater remediation system at Hanford. In a typical pump-and-treat system, like that featured in this illustration, water wells intercept a contaminant plume and pump water to the surface where it is treated to remove the contamination. Treated groundwater is usually returned to the aquifer. SOURCE: Department of Energy.

ter wells also provide a way to monitor progress of the remediation and to make adjustments to the system in response to changes in subsurface conditions.

In many circumstances pump-and-treat systems can be effective in preventing further migration of plumes, but they may be ineffective for eliminating them (Sidebar 2.2). Because aquifers do not have uniformly permeable strata (or uniform biogeochemical properties) pumping systems cannot uniformly remove contaminants. For example, in a sedimentary aquifer system comprising fine-grained silts interbedded with sands, most of the flow occurs in continuous sandy zones and little flow moves through the silts due to the permeability difference. Contaminants are flushed relatively quickly from the high-permeability zones while contamination in the low-permeability zones can remain largely in place. Once the high-permeability zones are flushed, the concentration gradient drives contaminants to diffuse from the low-permeability (high concentration) to the higher-permeability zones, where their concentration is diluted by greater flux.

SIDEBAR 2.2
Carbon Tetrachloride Plume Remediation at Hanford

As the preferred remedy identified in the draft feasibility study for the carbon tetrachloride (CCl_4) plume at the 200 West Area, Hanford recommended a pump-and-treat (P&T) system, already partially in place as part of an interim remedy, combined with flow path controls, institutional controls, and monitored natural attenuation. The purpose of the interim remedy for the ~11 km^2 CCl_4 plume (estimated area exceeding the water quality standard in the upper region of the unconfined aquifer) is to prevent further migration of the portion exceeding 2,000 μg/L. To date, the P&T remedy appears to be successful in achieving the stated goal.

From 1994 to 2006, the pump and treat system has removed approximately 10,000 kg of CCl_4 through the removal and treatment of over 3 billion liters of water using up to 10 extraction and 3-5 injection wells. For reference, the estimated mass of CCl_4 discharged to ground, mostly as a DNAPL oil mixture, was 577,000 to 922,000 kg. Additional extraction wells are planned.

Maps of the CCl_4 plume comparing the extent in 2006 to 1990 show that the 1,000-μg/L and 2,000-μg/L contours for the upper portion of the aquifer encompass smaller areas in 2006 compared with 1990. However, the portion of the plume at lower concentrations (<1,000 μg/L) has continued to expand.

Initial characterization of plume extent focused only on the surficial portion of the aquifer. Recent characterization has shown that CCl_4 is present at concentrations above the water quality standard throughout the thickness of the unconfined aquifer (~60 meters thick). High concentrations are found at depth and also displaced from the high concentrations at the surface. The impact of P&T to date on the deeper portion of the plume cannot be evaluated. The recent three-dimensional characterization has

Pump-and-treat systems in heterogeneous aquifers (as are most) typically produce a two-stage mass extraction behavior over time that has been attributed to permeability heterogeneity in the subsurface, as described above, in at least some cases. Rapid flushing of more permeable zones and high initial mass extraction rates are followed by tailing behavior typified by sustained contaminant concentrations (and mass removal) that change slowly. Differences in sorptive capacity among units or among grains of the geologic media can lead in a parallel way to variability in the capacity to store contaminants in the solid phase and similar types of behavior. The high concentration regions that remain within the groundwater system can cause the contaminant concentration to rebound when a pumping system is turned off. Concentration rebound following the cessation of pumping can also result from increased access to contaminant mass stored above the pumped water table in an unconfined aquifer.

resulted in a significant increase in the estimated CCl_4 mass present in the unconfined aquifer. The initial concentrations of CCl_4 present in the groundwater and the detections of CCl_4 throughout the aquifer thickness are both consistent with transport of DNAPL-containing CCl_4 oil below the water table.

Experience with this project points to scientific and technical challenges that are inherent in groundwater and soil remediation. The groundwater flow path to the Columbia River is very long, allowing for significant reaction between CCl_4 and the aquifer solids. In order to constrain risk predictions from a potential continuing source of CCl_4, characterization of the aquifer and groundwater biogeochemical and hydrogeologic system, as well as reactions with CCl_4, is needed. The uncertainty surrounding all aspects of the CCl_4 plume (presence of DNAPL, aquifer reactivity) was previously unknown, leading to extreme variability in risk assessment outcomes that drive the cleanup. Some recent activities surrounding the CCl_4 plume provide a positive model for reducing uncertainty, including improved characterization of the plume, for example. In addition, some research into CCl_4 reactions with aquifer solids, previously not evaluated, has recently been provided limited support by ERSP. However, the aquifer itself is relatively poorly characterized, especially with respect to the biogeochemical conditions that can impact the fate of CCl_4 and many other mobile groundwater contaminants relevant to EM at the field site, such as nitrate, uranium, and technetium. Similar to some other DOE sites in the western United States, the aquifer is deep and the geologic system is challenging to sample. Currently available methods do not provide robust information. There are new ERSP projects that seek to provide improved geologic characterization of contaminated sites through the use of geophysical tools. Improvement in methods of site characterization and monitoring in deep and/or challenging geologic systems offers the potential for significant benefit at DOE sites.

Impact of the Gap

Because of the heterogeneous distribution of physical and biogeochemical properties in groundwater, groundwater is inherently a poorly mixed system over short travel distances. As a consequence, pump-and-treat systems and other active remediation methods (methods that involve actively processing the groundwater in order to remove the contaminants) may be inefficient or ultimately ineffective. Operating these processes over the long time periods that may be required (e.g., 50-100 years) to meet cleanup goals can be expensive, and they may be discontinued or considered to have failed before the goals have been met. Premature implementation of pump and treat as the baseline technology can divert resources away from finding less expensive and ultimately more effective solutions.

Current Status

At Oak Ridge, DOE has used a continuous pump-and-treat system at the east end of Y-12 to keep an underground plume of carbon tetrachloride from spreading farther. Water is pumped to the surface, treated, and then released into a nearby creek. This is a large plume that is evidently being fed from an underground source of the carbon tetrachloride. Although it has been effective in limiting the plume's offsite migration, the treatment system has not eliminated the source or significantly reduced the concentration of carbon tetrachloride in the plume (Appendix F). A similar situation exists at Hanford, where pump and treat has controlled a carbon tetrachloride plume but impact on the source and plume longevity is unknown (Sidebar 2.2).

Whitaker (2008) described successful applications of two active remediation approaches at SRS that may be improvements over pump and treat. One is a steam injection and contaminant removal system that is remediating a 3-acre area regarded as the primary source of subsurface contamination in A- and M-Areas. This dynamic underground stripping system is expected to complete the remediation in 5 years versus an estimated 200+ years using conventional technologies. The system reportedly had removed 380,000 pounds of solvents at the time of the committee's visit. Second was the use of electrical resistance heating, which removed 710 pounds of solvents at C-Reactor in 2006. The system achieved 99 percent efficiency according to soil samples, and completed the cleanup 2 years faster than soil vapor extraction. In 2007, the aboveground equipment was relocated to an area referred to as the CMP pit where chemicals, metals, and pesticides were disposed (Whitaker 2008)

Although active remediation measures such as these at SRS can be effective, several NRC reports (1994, 1997b, 2000b) point to potential limitations of active remediation approaches. Three reports (NRC 1994, 1997b, 1999b) recommended that additional work be undertaken on *passive* remediation technologies. Passive barriers limit contaminant flux by reducing concentrations through biological or chemical reactions. As for any barrier system, barrier longevity is a critical aspect of success.

The In Situ Redox Manipulation (ISRM) barrier that was installed to remediate a chromium groundwater plume in Hanford's 100D Area is one case in point. Laboratory experiments performed before installation of the ISRM barrier indicated that it would be effective for approximately 20 years, but localized signs of failure were discovered after only 18 months. The cause of premature barrier breakdown was determined to be heterogeneities in the aquifer, where laterally discontinuous units with high permeability and lower inherent reductive capacity (because of lower iron content) were reoxidized faster than the less transmissive layers (DOE 2004a,b).

In the mid 1990s, OBER formed the Natural and Accelerated Bioremediation Research program to develop more fundamental insight into the interplay between geochemical and biological processes that may lead to effective and new bioremediation technology. This program consumed much of the OBER Subsurface Science Program and in 2005 was merged with the Environmental Management Science Program to create the ERSP, which supports fundamental, mission-oriented research on DOE legacy waste and priority contaminants. These research programs have generated nearly 1,000 research publications relative to the mechanistic microbiology, fate, and transport issues influencing metals and radionuclides in the subsurface and their potential for bioremediation and immobilization.

Approaches to Bridge the Gap

According to the committee's assessment of information it received, the following remediation approaches have promise for future EM R&D:

- Reactive chemical barriers,
- Natural attenuation, and
- Bioremediation.

Approaches such as natural attenuation (Sink et al. 2004) and PRBs and treatment zones[19] may provide EM with remediation solutions that are lower cost and more likely to succeed over the long term than baseline approaches such as pump and treat.

To enhance its use of MNA, EM needs a better technical basis for determining the situations for which it is an appropriate tool for reme-

[19] See http://clu-in.org/download/rtdf/2-prbperformance_web.pdf.

diation. This requires detailed understanding of contaminant associations, biogeochemistry, hydrology, and projections of future behavior. A recent EPA (2008) paper discusses site characterization to support use of MNA for remediation of inorganic contaminants in groundwater.

In the area of PRBs and treatment zones, research is needed on predicting and improving the lifetime of iron PRBs and the development and performance modeling/monitoring of non-iron-based systems. Several prior NRC reports provide information and recommendations associated with MNA and PRBs (NRC 1994, 1997b, 1999b, 2000b, 2007c, p. 15).

The concept of cocooning subsurface contamination described in GS-1 involves stabilizing contamination in place for now; monitoring it until radioactive decay, other natural processes, or new technologies make ultimate cleanup feasible or unnecessary; adapting to new knowledge, which may accumulate from R&D over years or decades; and making DOE's long-term responsibilities clear to all stakeholders.

Bioremediation approaches may provide cost-effective remediation options under conditions where other options are not feasible (e.g., aerobic, fractured bedrock at Oak Ridge, "tight" formations at SRS, and deep vadose zones at Hanford and Idaho). The challenge for bioremediation, however, is that most of the contaminants are either degraded slowly and/or incompletely, and they require use of growth-supporting substrates for *general* microorganisms. The organisms that efficiently degrade contaminants such as trichloroethylene, the predominant VOC at DOE sites, are very specific. Depending on whether the environment is aerobic or anaerobic, completely different organisms and biochemistries operate.

While many metals, minerals, and radionuclides may be biochemically oxidized or reduced by microorganisms, this capacity is again very specific to select groups of individuals or species. At Hanford, tests are under way to introduce lactate to promote subsurface anaerobic, hexavalent chromium reduction and immobilization. That work is being coupled to state-of-the-art gene expression monitoring to prove the in situ physiological basis of the process.

While new organisms that are capable of reducing contaminant concentrations have been found through DOE-supported research, they may not be predominant in the environment. Controlling their biochemical activity for actual application in groundwater remediation would be difficult without adequate fundamental understanding of the biogeochemical conditions of the specific area to be remediated and the ability to monitor the processes involved. Given the breadth of site conditions and contaminants of concern that DOE manages, key targets for remediation would be emphasized, such as the carbon tetrachloride plumes at Hanford and INL. These area-specific characterizations would be coupled to the development of methods for obtaining in situ biogeochemical information, fine-scale geophysical characterization, and high-resolution in situ monitoring. Subsequently, these investigations can improve the conceptual site models that are used to integrate the fine-scale structure, transport, and chemical reactivity that are needed to guide transport predictions and process optimization for site remediation.

Prioritization of the Gap

Relative to other science and technology gaps discussed in this section the committee judged the priority of addressing this gap as Medium.

	Relative Rating		
Criteria	High	Medium	Low
Volume of waste affected	X		
Potential to reduce technical uncertainty		X	
Potential to affect cleanup schedule		X	
Potential to affect cost			X

GS-3: The long-term performance of trench caps, liners, and reactive barriers cannot be assessed with current knowledge.

Engineered containment barriers, such as trench caps, liners, and reactive barriers, are designed to reduce risks associated with buried wastes and subsurface contamination by preventing the spread of contamination and/or minimizing the amount of surface- and/or groundwater that comes into contact with the wastes and contamination (Figure 2.7).

On the basis of as many as 20 years of observations, a recent NRC (2007a, p. 1) report that assessed the performance of engineered waste containment barriers concluded that "most engineered waste containment barrier systems that have been designed, constructed, operated, and maintained in accordance with current statutory regulations and requirements have thus far provided environmental protection at or above specified levels." The report (p. 2) also stated that although extrapolations of long-term performance can be made from existing data and models, such extrapolations will have "high uncertainties until field data are accumulated for longer periods, perhaps 100 years or more." In the case of natural ana logue systems, such as the Fernald mixed-waste landfill in Ohio, which are designed to serve 1,000 years or more without regular maintenance, the report (p. 73) concludes that "maintenance-free covers have not yet been demonstrated to work." Thus, the long-term performance of containment barriers, such as trench caps, liners, and lateral walls, cannot be assessed given the current state of knowledge.

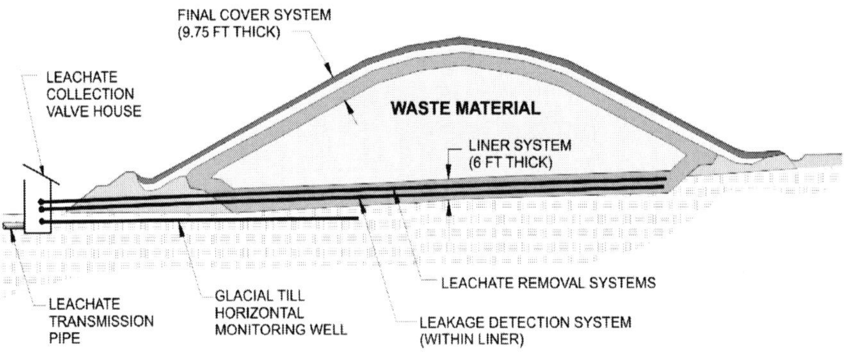

FIGURE 2.7 Diagram of an engineered, near-surface, waste disposal facility. Similar approaches, all of which depend on special materials and construction designs, are used for permanent disposal of low-level radioactive waste and hazardous materials at the DOE sites the committee visited. Such engineered facilities are intended to contain the waste for hundreds of years or more.
SOURCE: Department of Energy.

Impact of the Gap

Removal and subsequent treatment of wastes at many DOE sites is technically difficult, expensive, and potentially hazardous to workers. As a result, alternative approaches that leave the waste in place, but incorporate robust containment barriers and waste stabilization technologies over the long term (100-1,000 years), are a key element of DOE's strategy for managing legacy waste sites. Without confidence in predicting the performance of containment barriers beyond a few decades, many of DOE's performance assessments, which assume long-term barrier integrity, could be deemed unreliable enough to prevent current plans for area and site closures going forward. Repair or replacement of engineered barrier systems that fail in a relatively short time could increase costs, delay site closure, and raise stakeholder concerns about the likelihood of future failures.

Current Status

The behavior of engineered barrier systems is influenced by environmental and ecological conditions that will continue to evolve over time as a result of processes like ecological and biological successions, landform evolution, and climate change. Research in the INL Environmental Surveillance, Education and Research Program[20] and at the Savannah River

[20] See: http://www.stoller-eser.com/research.htm.

Ecology Laboratory (SREL)[21] is aimed at understanding such processes, thus providing information relevant to the robust design, construction, and maintenance of containment barriers.

A Hanford-designed prototype surface barrier, referred to as the Hanford barrier, is a 2.5-hectare multilayered, vegetated, capillary barrier composed mainly of stable natural materials and designed to isolate buried wastes for about 1,000 years. While not all near-surface disposals at Hanford will require the degree of protection offered by the Hanford Barrier, Ward (2007) stated that the results of tests and monitoring of the barrier's performance can be used to guide the design of more modest covers. To do so, it will be necessary to determine moisture flux through representative waste sites, including vegetated and graveled surfaces, account for seasonal variations in precipitation and heating, and from this information develop robust infiltration barriers for sites where contaminants will be left in place (Appendix D).

The Alternative Cover Assessment Program (ACAP) is developing field-scale performance data for landfill final cover systems based on field data being obtained at a dozen sites representing a variety of geohydrologic conditions.[22] ACAP is part of the EPA's National Risk Management Research Laboratory's SITE program established to promote the development of new and innovative technologies used to address hazardous waste problems. Both prescriptive (Resource Conservation and Recovery Act) and innovative alternative cover designs are currently being tested in the project.

SREL is investigating the alternative use of native grasses for vegetated caps (Kwit and Collins 2008), while projects at INL are exploring evapotranspiration cap designs (ET barriers) as a low-cost, low-maintenance alternative to traditional designs.[23] EM's Office of Engineering and Technology recently hosted a workshop on landfills (Benson et al. 2008).

Approaches to Bridge the Gap

According to this committee's assessment of information it received, the following approaches have promise for future EM R&D:

1. Monitoring systems that can provide information on containment barrier performance can (i) reduce uncertainty related to the long-term performance of such engineered controls and (ii) lay the foundation for approaches that can provide early warning of unexpected or unacceptable barrier behavior so repairs or adjustments can be made in a timely fash-

---

[21] See: http://www.uga.edu/~srel/.

[22] See http://www.acap.dri.edu/.

[23] See: http://www.stoller-eser.com/NERP/PCBE.htm.

ion. Monitoring systems might include buried sensors, surface or airborne surveillance, eco- and/or bio-indicators, and software support. Robust and low-cost systems that reduce manual labor and allow for remote, real-time access to data on barrier performance offer the most promise for improved monitoring.

2. Robust models of barrier behavior that can incorporate appropriate uncertainty and account for natural and anthropogenic spatial and temporal changes, together with field data to calibrate these models, can better assess long-term barrier behavior. To identify unacceptable barrier behavior, a scientific basis for what is an unacceptable "barrier breach" would be needed.

3. Many of the barrier systems put in place or proposed at DOE sites are systems that are designed to shed precipitation and/or divert or retard groundwater flow. Thus, they are systems that are intended to resist natural processes rather than work with them. These systems cannot be expected to provide long-term waste or contaminant isolation without continued maintenance or, in some cases, replacement and remediation at considerable effort and cost. The continued development and performance monitoring of alternative systems, such as natural analogues to existing landscapes and ET barriers that can work with rather than against nature, are thus needed (Clarke et al. 2004).

4. Engineered barriers, including those described in approach 3, above, might be better recognized as temporizing measures to control contaminant spread for years or decades until new technologies or natural processes provide a final solution. According to the cocooning concept introduced in GS-1, R&D would first be directed at ensuring safety of the engineered barrier system, with continued R&D toward improved characterization, monitoring, and modeling to ensure safe, permanent disposition of the contaminants or the contaminated area.

5. For the barrier designs themselves, R&D could be directed at developing a design strategy, and associated life-cycle cost model, that involves "perpetual periodic replacement" of the cover or barrier system rather than a design philosophy based on barriers that need to last as long as practicable.

Prioritization of the Gap

Relative to other science and technology gaps discussed in this section the committee judged the priority of addressing this gap as Medium.

Criteria	Relative Rating		
	High	Medium	Low
Volume of waste affected		X	
Potential to reduce technical uncertainty	X		
Potential to affect cleanup schedule			X
Potential to affect cost		X	

GS-4: The long-term ability of cementitious materials to isolate wastes is not demonstrated.

Very large volumes of cementitious materials will be used in the EM cleanup with the objective of protecting soil and groundwater by encapsulating wastes. Cementitious materials are among the world's most widely used and best understood construction materials. In the EM cleanup program, their high-volume applications include:

- Grouting of emptied HLW tanks and associated inter-tank transfer pipes;
- Stabilizing LAW in large monoliths, such as the SRS saltstone, or in smaller containers (e.g., 55-gallon drums); and
- Constructing disposal vaults or other structures.

Cementitious materials are the best, and as a practical matter the only ones, available for these applications. However, ensuring that they can effectively isolate waste for hundreds of years or more will be an ongoing scientific and technical challenge.

Before the HLW tanks at Hanford, Idaho, and SRS can be closed, they are to be filled with a cementitious grout material, which has two purposes: (1) to encapsulate or otherwise reduce the mobility of the residual waste and (2) to stabilize the tank structurally to support the overburden. Hanford has 177 tanks, Savannah River 51 tanks (2 of which have been filled with grout), and Idaho 11 tanks (8 of which have been filled with grout). These are large, typically 100,000- to million-gallon tanks, from which it is not possible to remove all of the contaminated material in the tanks; see WP-1. Stabilizing the residual waste in these tanks is essential.

Currently there is almost no experience in cleaning and grouting inter-tank transfer pipes that were used to move waste among the waste tanks. Relative to their volume, some of the pipes, especially those that have become plugged, are likely to contain more residual wastes than the tanks and be considerably more difficult to fill thoroughly with grout.

Salt waste in the tanks at SRS is to be processed and the resulting

LAW incorporated into a grout called saltstone; see WP-2. Depending on the specific constituents of the salt solution, the grout is formulated using appropriate proportions of portland cement, fly ash, ground granulated blast-furnace slag, water, and chemical admixtures. The grout is pumped into concrete vaults, where it hardens. Saltstone has a low oxidation-reduction potential (E_h) to stabilize key radionuclides such as Tc-99 in less soluble forms to reduce the rate at which they would leach out or migrate in the groundwater (Rosenberger et al. 2005; Shuh et al. 2002). The saltstone vault has a concrete roof and will eventually have an engineered cap over the entire installation (NRC 2006b). The vault is also a barrier to contaminant release.

Impact of the Gap

The successful grouting of wastes in tanks, pipes, and saltstone is assumed in performance assessments that demonstrate regulatory requirements for tank closure and SRS salt disposal will be met. If the adequate long-term performance of the grout were to be seriously questioned—and the requirements for the grout's performance and performance period are beyond any direct experience in the construction industry—then closure of the tanks and SRS salt disposal could become problematical.

The Nuclear Regulatory Commission (USNRC 2005) Technical Evaluation Report (TER) on the DOE's performance assessment of salt waste disposal at SRS (p. 50) indicates that the Commission has concerns about uncertainties in saltstone's performance:

> In conducting its PA [performance assessment] of the facility, DOE considered the various mechanisms of release to estimate the source term and release of contaminants. Both diffusive and advective transport processes were addressed. To model contaminant transport in the near field, there was a need to estimate the contaminant concentrations in the pore fluid based on the concentrations in the saltstone. However, relating inventory in the saltstone to pore fluid concentration is complicated by various processes, such as precipitation/dissolution reactions, aqueous complex formation, and sorption. DOE acknowledged that these processes are poorly understood and difficult to quantify for the SDF [saltstone disposal facility].

The TER (p. 52) summarizes the USNRC's evaluation of the DOE's model of saltstone and concrete vault degradation as follows:

> In general, the NRC staff agrees with the qualitative assessment of the degradation mechanisms for saltstone. However, given that: (1) the calculated releases from the SDF are sensitive to the values of hydraulic conductivity of the vault and saltstone; and (2) "the timing and extent of degradation

> are not readily predictable due to enormous uncertainties in conditions for thousands of years" (Cook and Fowler, 1992, Section 3.1.3.5), it would be useful to reduce the uncertainties associated with the hydraulic conductivity and long-term integrity of the vault and saltstone. *Additional laboratory measurements of initial hydraulic conductivity, as well as long-term tests or monitoring studies designed to evaluate the long-term durability of the saltstone and concrete vault, would help reduce these uncertainties* [italics added].

The TER also summarizes the factors that are important in assessing compliance with 10 CFR Part 61, Subpart C. It notes on page 90, "some of the assumptions made in the analysis, if incorrect, could lead to noncompliance with the performance objectives."

Current Status

Cementitious grouts and related materials are routinely used in the construction industry for a wide variety of applications, some of which closely match EM's needs. Where the project requirements are the same as or similar to these routine applications, the DOE can simply use existing technology. For example, controlled low-strength materials are used in bulk to fill utility trenches and provide some load-carrying capacity, which would be similar to bulk-filling a large waste tank to provide structural stability. However, there are some requirements that are unique to DOE applications:

- Grout mixtures must be suitable for pumping into the tanks, typically through "tremies" (long, movable pipes that allow placement of the grout into specified locations in the tank (Figure 2.8) without the components of the grout (mainly cement, sand, and water) separating;
- They must provide near- and long-term chemical conditions (high pH and low E_h) to maintain the radionuclides and toxic heavy metals in their least mobile forms; and
- They must minimize the flow of water through the material (and the consequent leaching of radionuclides and metals from the grout) (NRC 2006b).

In recent years the construction industry has begun to build structures with design lives of 75 to 100 years based on a combination of experience with concrete structures over decades and modeling. However, the design lives of grouted DOE wastes are intended to be 10 times longer or more. The short-term behavior of radionuclides and toxic heavy metals in waste-form grouts is reasonably well understood, but the long-term behavior is

FIGURE 2.8 Tremie being used to emplace grout in a tank at the INL site. The use of tremies (long flexible pipes) is common in the construction industry. For EM's applications, the components of specially formulated grout mixtures must not separate before they reach their in-tank destination and solidify.
SOURCE: Department of Energy.

not. For example, it is well known that the pH of cementitious materials decreases over time due to carbonation (reaction with carbon dioxide from the air), while little is known about how other properties such as E_h change with time. The models currently being used to predict long-term performance of tank grouts and saltstone necessarily extrapolate from very limited and relatively short-term data.

Concrete vaults are constructed to contain SRS saltstone. The vault wall, floor, and ceiling are part of the engineered barrier system expected to reduce the ingress and egress of water. Some of the vaults have already cracked and may be transmitting water, but the causes and potential fixes are not yet understood (USNRC 2008).

The USNRC outlined several cooperative research efforts focused on the long-term behavior of cementitious systems (Kock 2008). The DOE cement consortium includes both a simulation component and an experimental component. It is led by the DOE and includes the USNRC, Vanderbilt

University, the National Institute of Standards and Technology, SIMCO, the Netherlands Energy Research Center, and SRNL. It is funded by the USNRC Office of Research. In addition, the USNRC is sponsoring contractor research on cement-based materials, including degradation mechanisms, modeling, fast pathways, hydraulic conductivity studies, and a test bed. The USNRC Office of Research is also examining test methods, designs, additives, and monitoring techniques.

Approaches to Bridge the Gap

According to this committee's assessment of information it received, the following approaches have promise to lead to improved understanding of the long-term performance of cementitious materials and, thus, to improvements in the materials per se:

1. Improved data to support performance assessment models. The models currently being used to predict long-term performance necessarily extrapolate from very limited short-term data. Current models are believed but not known to be conservative. Monitoring the near- and long-term performance in the field would greatly improve the accuracy of the models and allow for adjustments to the grout formulations for future tank closures. The USNRC TER (p. 78) provides an example of how empirical data from the field could be used for this purpose:

> One of the key elements of DOE's PA is the [chemical] reduction of Tc-99 in the wasteform by the addition of slag. As previously discussed, the sensitivity analyses demonstrate quite clearly that the rate and extent of oxidation of the wasteform is a key factor in meeting the protection of the public performance objective. DOE has performed basic research to evaluate whether the slag would result in Tc-99 being contained in a reduced form, and installed field-scale saltstone lysimeter tests with and without slag (Cook and Fowler, 1992). . . . Currently, DOE's estimates for the amount of oxidation of the saltstone over 10,000 years are based primarily on numerical modeling results. *It may be possible to exhume and characterize a saltstone lysimeter. The depth of the penetration of the oxidation front should be able to be estimated and it would provide excellent model support for a key element of DOE's PA* [italics added].

The USNRC TER also noted inconsistent results for the measured hydraulic conductivities (permeabilities) of saltstone samples made or tested under different conditions. Very low conductivity must be achieved and maintained for saltstone to meet its performance requirements. Developing improved and quality-assured methods to measure hydraulic conductivities of very low permeability materials can assist the SRS tank closure program.

2. Basic understanding of the chemistry for improving the long-term performance of cementitious waste forms. In addition to limiting access of water to the waste, cement-based materials are expected to maintain the wastes in a chemically reducing environment to reduce the solubility of waste components. Properties such as reducing capability are not relevant in the construction industry; therefore, fundamental research, along with gathering the empirical data described in the USNRC TER above, are specific opportunities for EM to address this knowledge gap.

Pore fluids in grouted wastes can concentrate and release contaminants. However, relating the waste inventory in the saltstone to pore fluid concentration is complicated by various processes, such as precipitation/dissolution reactions, aqueous complex formation, and sorption. These processes are poorly understood and difficult to quantify, but they must be understood in order to ensure that grouted wastes meet long-term performance objectives.

3. Technology development for pipe grouting. The various tank farms have underground piping to carry wastes from one tank to another. Excavating these pipes and fittings is not practical, so they will be left in place. They will most likely be grouted with a material similar to that used on the bottoms of the tanks, that is, an engineered grout with reducing properties. This material will need to flow into place under pressure, fill or nearly fill the pipe, and not set until it has done so. Once the material sets, there will be no possibility of pumping additional material into or through the pipe.

For this application, the shrinkage of the grout must be controlled. All cementitious materials shrink to some degree on hydration, which would leave the pipe less than completely filled. Materials can be engineered to shrink less, and shrinkage-compensating materials can also be designed. These formulations expand first and then shrink. When the initial expansion is restricted, as it would be inside a pipe, it could induce radial (bursting) stresses in the pipe. Learning how to better control expansion or shrinkage of cementitious materials in confined areas is an opportunity for EM to partner with industry.

4. Improved understanding of crack formation and mitigation in concrete vaults. Cracking of construction concretes is a well-known phenomenon. Many of the causes for cracking and the means for mitigation are well understood. However, in construction, cracks are expected and tolerated within certain limits, and cracks exceeding these limits can often be satisfactorily repaired. New methods of design, detailing, and construction may be required where tighter tolerances or extended durability are necessary. New repair materials and methods suitable for areas with limited accessibility could also be helpful.

Prioritization of the Gap

Relative to other science and technology gaps discussed in this section the committee judged the priority of addressing this gap as High.

	Relative Rating		
Criteria	High	Medium	Low
Volume of waste affected	X		
Potential to reduce technical uncertainty	X		
Potential to affect cleanup schedule	X		
Potential to affect cost	X		

DEACTIVATION AND DECOMMISSIONING

The deactivation and decommissioning and facility engineering program area of the EM roadmap deals primarily with facility characterization; deactivation, decommissioning, and demolition; and closure.[24] Principal science and technology gaps the committee identified in this program area are described in this section.

Facilities requiring D&D throughout the DOE complex include reprocessing plants, large production and smaller test reactors, fuel fabrication facilities, gaseous diffusion plants, and laboratories with hot cells—including all of these facilities' support structures that typically contain ancillary equipment, piping, and ductwork. In many cases there are complicating factors including poor (and continually degrading) condition of structures, associated chemical hazards, and nearby active facilities with ongoing operations.

While hundreds of DOE facilities have undergone D&D, some 3,000 remain to be decontaminated and removed or closed, including many of the most challenging ones.[25] A previous report (NRC 2001c) noted that cleanup of facilities will be technically challenging due to (i) personnel hazards; (ii) large size of facilities, including those with massive shielding structures; (iii) complex, crowded and often retrofitted arrangement of equipment and support structures; (iv) poorly understood and difficult-to-characterize contaminants; and (v) lack of decisions on end states. Many buildings and facilities are to be partially or completely demolished while some massive structures will be decontaminated and left in place. The contaminants to be removed include solids and liquids that can be radioactive,

[24] For convenience this report will use the abbreviation D&D in referring to EM work in this program area.

[25] See http://www.em.doe.gov/Pages/BudgetPerformance.aspx.

chemically hazardous, or both. Some contaminants may be easy to remove and others strongly bound to a substrate (e.g., concrete, steel).

According to information gathered by the committee, the most difficult D&D challenges include radiochemical separation facilities at Hanford, Idaho, and SRS; production reactors at SRS; gaseous diffusion plants at Oak Ridge, Paducah, and Portsmouth, plutonium processing plants at Hanford, Los Alamos, and SRS; tritium processing facilities at SRS (NRC 2001c), and support facilities (including sewage lines) at SRS (Whitaker 2008).

DD-1: D&D work relies on manual labor for facility characterization, equipment removal, and dismantlement.

Currently D&D projects require extensive hands-on, manual labor that unavoidably exposes workers to hazardous conditions (Figure 2.9). Besides the rather obvious hazards to workers who manually dismantle, size reduce (cut up), and remove contaminated structures and equipment, each facility requires extensive characterizations to determine the nature of contaminants before, during, and after D&D. Characterization exposes workers to radiation and other hazards and is costly, amounting to some 15 to 25 percent of overall D&D budgets (NRC 2001c). Work must sometimes be done in high-radiation environments. For example, at Idaho a technical challenge is to characterize and remove contamination in pipelines and other structures that have high-radiation fields (up to 1,600 rads/hour) and are located under a building at the site (NRC 2007c, p. 28).

Workshop panelists representing Oak Ridge agreed that D&D is a top priority for the site, mainly due to challenges presented by the gaseous diffusion plants (manual removal of transite siding from these very large buildings was cited (Figure 2.10) and other deteriorating structures (NRC 2007c; McCracken 2007). SRS D&D priorities are worker protection and characterization of facility "hot spots" (NRC 2007c, p. 27).

Impact of the Gap

Safety of workers and of the public is the primary consideration in the EM cleanup. Worker safety is a criterion for contractors. Should an incident occur that harms a worker or could have caused harm, operations are halted until the incident has been thoroughly investigated, the cause is determined, and measures to prevent such future incidents are implemented. No matter how carefully planned and carried out, hands-on D&D work carries a high risk for radiation exposure, bodily uptake of radioactive or hazards materials, and injury.

FIGURE 2.9 Hands-on D&D work. Facility D&D often requires hands-on work with large, contaminated equipment in hot, confined spaces. Although uncomfortable, personal protective equipment like that worn by the worker in this photograph is necessary to protect workers from the uptake (skin, mouth, nose) of radioactive or other hazardous substances and from physical hazards.
SOURCE: Department of Energy.

FIGURE 2.10 Transite removal at Oak Ridge. Transite was a commonly used siding material throughout the DOE complex. Today's workers must wear personal protective equipment and follow special procedures to remove this asbestos-containing siding. Transite is heavy and often has to be handled in confined, elevated work spaces as shown here.
SOURCE: Department of Energy.

Current Status

Manual labor has been key to EM's D&D work, including the successful closure of the Rocky Flats site under budget and ahead of schedule. Rocky Flats was formerly a major plutonium-handling site, which has now been converted into a wildlife refuge—a major accomplishment. Hands-on labor for D&D is a good example of the committee's considering technology gaps as potholes in a road that EM can work around. EM and its contractors can and have managed worker safety for hands-on D&D. Nonetheless, R&D toward removing workers from a hazardous environment could provide a better solution.

Robotics and remote manipulation for sensing, inspection, measurement, and tank waste remediation have been developed and deployed to some extent at both the Savannah River and Hanford sites. DOE has made limited use of some robotic technology as part of the Glovebox Excavator Method used to demonstrate retrieval of buried TRU waste at Idaho (NRC 2005, p. 43).

Researchers at INL have been exploring the possibility of using semi-autonomous robotic systems for detection and characterization in radiological environments. These systems may reduce some uncertainties inherent in different training and skill levels among operators while allowing tasks to be completed more quickly than in the case of purely teleoperated systems (Nielsen et al. 2008). In all cases, the purpose of employing robotic and remote systems is to reduce D&D worker risks while accelerating the pace and accuracy of the remediation operation.

SRNL is extending its previous experience with remote devices for use in radiation areas to develop robotic and teleoperated systems for homeland security and defense applications. Non-DOE agencies and universities, including the National Aeronautics and Space Administration (NASA), National Oceanic and Atmospheric Administration, and Carnegie Mellon University, conduct research on robotics and remote-operator systems for the Department of Defense and for ocean and extraterrestrial exploration.

There are also recent efforts outside the United States to develop robotics and remote systems for decommissioning of former nuclear power facilities. For example, a group at Lancaster University in the United Kingdom has been funded by the Nuclear Decommissioning Authority to develop a multiarmed robotics system that would allow D&D operations in the United Kingdom to be faster, safer, and more cost-effective, and reduce the radioactivity dose levels to which workers are exposed (Bakari et al. 2007). Work at the French Atomic Energy Agency and COGEMA has focused on radiation-hardened electronics and force feedback mechanisms used in telerobotics operations involving spent fuel (Desbats et al. 2004).

Approaches to Bridge the Gap

According to this committee's assessment of information it received, the following have promise for future EM R&D:

1. Improved technologies that could reduce worker exposure by reducing the need for manual sample collection. These include:

- Devices for rapid characterization of low levels of contamination (radionuclides and EPA-listed substances) on surfaces of construction materials and equipment, including devices that can detect very-low-energy beta emitters (e.g., tritium), low-energy photon emitters (iodine-129), and beryllium;
- Minimally invasive methods to characterize contaminant concentrations as a function of depth in construction materials, especially concrete; and
- Instruments for remote mapping of radionuclide contamination at low levels that can differentiate specific radionuclides, including beta and alpha emitters.

2. Greater use of robotics to reduce manual labor and worker risks. NRC (2002) recommended that DOE develop robotic technologies for retrieval and repackaging of buried waste. NRC (2001c) recommended research to develop intelligent and adaptable robotic systems that can be used for facility decommissioning. Next-generation robotic systems will need to be:

- Adaptable to a variety of environments and topographies;
- Semi-autonomous to provide a more intuitive human-robot interface, prevent accidents, and optimize execution of tasks; and
- Highly reliable.

Such needs were recognized in EM's former D&D Focus Area 10 years ago (Staubly and Kothari 1998) and remain at the forefront of R&D in robotics.

Prioritization of the Gap

Relative to other science and technology gaps discussed in this section the committee judged the priority of addressing this gap as High.

Criteria	Relative Rating		
	High	Medium	Low
Volume of waste affected		X	
Potential to reduce technical uncertainty[a]	X		
Potential to affect cleanup schedule	X		
Potential to affect cost	X		

[a]Including risks to workers.

DD 2: Personal protective equipment tends to be heavy and hot and limits movement of workers.

As described in DD-1, manual D&D work at all sites requires workers to perform safely and efficiently in hazardous environments. Broadly speaking, personal protective equipment (PPE) can range from standard items such as coveralls, safety glasses, and gloves, to face masks with capability to filter or detoxify airborne contamination ("assault masks"), to full-body anti-contamination suits for work in heavily contaminated areas (Figure 2.9). Anticontamination suits encapsulate the entire body in an impervious suit, and provide safe breathing air by means such as filtration of ambient air, use of self-contained breathing apparatus, or an external supply of uncontaminated air delivered through a flexible hose.

PPE for less-contaminated workspaces consists of some type of protective clothing, often in multiple layers, which encloses most or all of the body. PPE is often heavy and bulky, resulting in limitation of motion, extra exertion, and overheating with the consequent risk of heat stress (Bernard 1999). Protective clothing that does not allow perspiration to escape increases body temperature, which reduces worker comfort and productivity (DOE 1998b).

Impact of the Gap

The limitation of motion and extra exertion imposed by PPE required in high-contamination zones can cause worker stress and reduce the efficiency of D&D work. PPE with externally supplied cool air can reduce heat stress but can have various limitations and problems related to the supply hose. During its Idaho visit, the committee was shown a waste retrieval operation in the Radioactive Waste Management Complex in which workers can operate excavation equipment for only short periods of time due to the risk of heat stress. This, coupled with the time to don and doff PPE, increases the duration and cost of D&D activities.

Current Status

PPE is used throughout the nuclear and hazardous materials industries. There are companies that develop and manufacture PPE (see, e.g., Frham Safety[26] and G/O Corporation[27]). EPRI and USNRC are supporting technology for improving PPE. Shedrow (2008) noted that SRNL has developed a variety of PPE technologies that have been used in environmental remediation work at SRS and other locations.

Approaches to Bridge the Gap

According to this committee's assessment of information it received, there is a need for PPE designed for elevated temperatures and longer exposures in contaminated environments. Lighter and cooler PPE would allow workers to safely remain longer in the presence of hazardous materials. There are opportunities to adapt available technologies (e.g., from NASA, Department of Defense). For example, adaptations of NASA protective clothing technology have been examined for use in development of protective clothing for firefighters (Foley et al. 1999). The Department of Defense has supported a number of programs for development of advanced impermeable "NBC" (nuclear/biological/chemical) anticontamination clothing for a number of years, citing this area of need in the Defense Technology Area Plan (DOD 1999). This technology has not been adapted and adopted in D&D applications. Further evaluation would seem appropriate.

Robotics and remote or teleoperated techniques will also limit worker exposure, although there are circumstances (i.e., inspection, removal in very complex areas, sensitive structures) where manual labor is essential.

Prioritization of the Gap

Relative to other science and technology gaps discussed in this section the committee judged the priority of addressing this gap as Low.

	Relative Rating		
Criteria	High	Medium	Low
Volume of waste affected			X
Potential to reduce technical uncertainty[a]		X	
Potential to affect cleanup schedule		X	
Potential to affect cost			X

[a]Including risks to workers in this instance.

[26]http://frhamsafety.com/anti-c/encapsulating_suit.htm.
[27]http://www.gocorp.com/.

DD-3: Removing contamination from building walls, other surfaces, and equipment can be slow and ineffective.

Decontamination of facilities and equipment is carried out at multiple stages of the decommissioning process in order to lower worker exposure, prepare equipment for disassembly and removal, and prepare a facility for tear-down and removal (to limit release of contaminants prior to further treatment and disposition of the debris). A primary objective of decontamination procedures is to generate a small volume of the most hazardous waste, while the larger volumes of waste have low or no hazard, thus reducing the cost and long-term risk of their disposal. Some decontaminated equipment or facilities might be recycled or reused. The end state of any decontamination activity must be consistent with both site-specific and overall DOE cleanup objectives.

Concrete, such as that in the large canyon buildings on the SRS and Hanford sites and reactor shielding structures at multiple DOE sites, constitutes most of the volume and weight (estimated at over 27 million tons) of DOE's surplus facilities. Because of its inherent porosity, its heterogeneous surface structure (pits, cracks, and smooth and rough areas on both the macro- and microscopic scales), and its chemistry, concrete poses special challenges for decontamination.

At present, the usual method for removing surface contamination is called "scabbling"—the physical removal of the surface by workers in protective clothing using power tools. This procedure generates a great deal of dust and is hazardous to workers. Because of long-term exposure, the concrete is often contaminated to a depth of several millimeters beneath its surface (DOE 2000), and in some cases, such as for tritium, considerably deeper. In many instances, paints, sealers, and varnishes on concrete surfaces create a laminate problem, with aged materials being harder to decontaminate than more recent deposition (NRC 2001c).

Contaminated equipment including glove boxes, shielded cell liners, lead shielding, and plastic parts, along with heavily corroded surfaces, pose particular problems due to geometries and occluded structures that trap contaminants. In addition, effectiveness of D&D methodologies can be severely compromised due to the inherent difficulty of characterizing both the chemical nature of contaminants and the degree of their removal following decontamination resulting from occluded, porous, and heterogeneous surfaces of degraded structural building materials (Halada 2006). Before, during, and after the process of decontamination, it is necessary to identify contaminants on concrete and other structural surfaces. Nondestructive methods would be far preferable to the physical removal of samples (e.g., cement cores, metal coupons) for analysis.

Impact of the Gap

Current decontamination processes used by D&D contractors are labor-intensive and costly, and there is the ever-present risk of exposure to toxic and radioactive materials; see DD-1. These processes also generate large volumes of contaminated secondary wastes and often leave behind unwanted residual contamination. The risk of accidents is increased by the bulky protective clothing; see DD-2. Because of cost and hazards, cleanup contractors often choose to dispose of contaminated equipment and construction materials as wastes rather than to decontaminate and recycle them. While current baseline decontamination technologies probably can be made to work for future D&D work, there are opportunities to do the job more safely and cheaply and achieve higher degrees of decontamination by developing and using new technologies.

Current Status

The EPA has recently conducted two workshops on decontamination methods for chemical, radiological, and biological contaminants through its Office of Research and Development's National Homeland Security Center (EPA 2005, 2006a). In addition, the EPA has developed a reference guide, "The Technology Reference Guide for Radiologically Contaminated Surfaces," which provides a broad overview of chemical and physical methods for removing contamination from surfaces (EPA 2006b). These surveys and associated reports consider a broad range of options for decontamination. The technological challenges considered in the EPA report have much in common with DOE site needs, including a need for faster and more effective decontamination methods, determining surface chemistry interactions, difficulties with vertical surfaces and reaching high work areas with decontamination equipment, decontamination of tiny cracks and seemingly inaccessible areas, subsurface effects, and waste generation.

Investigators at INL completed a comprehensive study of removal and collection of radioactive contamination from building exteriors, which was supported by the Defense Advanced Research Projects Agency (Demmer at al. 2007). Activities in the United Kingdom and Canada are also of interest. For example, the effect of weathering and other environmental conditions on the association of radiological contamination with porous surfaces and resulting implications for decontamination have been considered in research by the Chemical, Biological, Radiological-Nuclear and Explosives Research and Technology Initiative Secretariat of the Defence Research and Development Canada, Centre for Security Science.[28]

[28] See http://www.css.drdc-rddc.gc.ca/crti/invest/rd-drt/02_0067rd-eng.asp.

Approaches to Bridge the Gap

Scientific understanding of the interactions among contaminants and construction materials is fundamental to developing more effective D&D technologies. Such information includes how contaminants bind to steel and concrete surfaces; how they penetrate into these materials; their migration into pores, fissures, and welds; and time-dependent aging effects. NRC (2001c) identified decontamination as an important science and technology gap and recommended specific areas of research needed to improve decontamination technologies, including:

- Development of a fundamental understanding of the chemical and physical interactions of important contaminants with the primary materials of interest in D&D projects, including concrete, stainless steel, paints, and strippable coatings to gain a better understanding of how contaminants bind to and penetrate these materials. This would involve understanding the interactions both kinetically and thermodynamically under a variety of conditions (pH, temperature, ionic strength);
- Development of dry decontamination technologies, including use of supercritical fluids such as carbon dioxide, that can be used to remove high levels of contamination with minimal secondary wastes (Appendix D);
- Exploration of the role of nanotechnology (for more efficient chelating) and biological mechanisms (including bioleaching, biosurfactants, biocatalysis, and cell-less enzymatic processes) for more efficient and rapid decontamination methods;
- Advanced methods to leach/migrate contaminants from cementitious matrices (Appendix D); and
- Development of decision tools for determining optimal decontamination approaches.

Prioritization of the Gap

Relative to other science and technology gaps discussed in this section the committee judged the priority of addressing this gap as Medium.

	Relative Rating		
Criteria	High	Medium	Low
Volume of waste affected		X	
Potential to reduce technical uncertainty		X	
Potential to affect cleanup schedule		X	
Potential to affect cost		X	

CONCLUSIONS

This chapter has presented 13 gaps that the committee views as the principal impediments to the EM site cleanup program. They are obstacles or impediments in the sense that they can represent likely causes for schedule delays, cost increases, and potential failures to meet currently envisaged cleanup objectives. Developed through the committee's site visits and other information gathering, all of these gaps are worthy of EM's consideration in developing future science and technology roadmaps. The committee was mindful of the research initiatives set forth in the EM roadmap but has provided its own independent assessments in this chapter.

The committee's prioritization of these gaps, given in Table 2.1, reflects a variety of technical judgments, including schedule and budget impacts, risk reduction, and likelihood of new technology developments that can bridge the gap. The committee has not attempted to be prescriptive by recommending specific research to address each gap, but rather it has indicated R&D approaches that it judges are most likely to bear fruit.

The committee used this chapter as a basis for developing the remainder of this report. Chapter 3 describes the personnel expertise and physical infrastructure that EM will need to carry out this R&D. Chapter 4 describes approaches and opportunities for EM to leverage its R&D work with other organizations.

3

Expertise and Infrastructure

This chapter addresses the third and fourth items in the committee's statement of task, which asked the committee to identify:

- Core capabilities at the national laboratories that will be needed to address the Office of Environmental Management's (EM's) long-term, high-risk cleanup challenges, especially at the four laboratories located at the large Department of Energy (DOE) sites (Idaho National Laboratory [INL], Oak Ridge National Laboratory [ORNL], Pacific Northwest National Laboratory [PNNL], and Savannah River National Laboratory [SRNL]).
- The infrastructure at these national laboratories and at EM sites that should be maintained to support research, development, and bench- and pilot-scale demonstrations of technologies for the EM cleanup program, especially in radiochemistry.

To address its task statement the committee interpreted the term "core capabilities" in the first task item above to refer to the scientific and technical expertise of personnel at the national laboratories. The term "infrastructure" was taken to refer to physical facilities (i.e., buildings and equipment). Because scientific and technical personnel require appropriate facilities with which to conduct their work, and physical facilities are useless without appropriately skilled personnel to operate them, the committee chose to address both task items together in this chapter.

The term "capabilities" will be used throughout this chapter to refer to both physical facilities and the personnel who have the needs and skills to use them. When distinctions need to be made, the more specific terms—expertise or infrastructure—will be used.

After reviewing the science and technology gaps identified in Chapter 2, the committee identified four kinds of capabilities (expertise and infrastructure) that will need to be maintained. Selecting these capabilities was a two-step process. First, a list of potential capabilities that would need to be maintained was developed from information gathered during the site visits and from committee members' expertise and experience. Then this list was culled based on consideration of which capabilities were essentially unique to the DOE cleanup situation. The result was the following four:

- Handling radioactive materials,
- Conducting engineering and pilot-scale tests,
- Determining contaminant behavior in the environment, and
- Utilizing state-of-the-art science to develop advanced technologies.

The remainder of this chapter describes each of these capabilities and their relevance to sustaining EM's future research and development (R&D) programs.

HANDLING RADIOACTIVE MATERIALS

The capability to work with radioactive materials is fundamental to EM's engineering and technology development. At least some stages of the R&D to address each of the gaps identified in Chapter 2, with the possible exceptions of DD-1 and DD-2, will require the use of radioactive tracers or actual radioactive waste.

All of the national laboratories visited by the committee have the facilities and personnel for handling radioactive materials. All of their nuclear-related initiatives (e.g., energy, defense, medicine) require this capability. This capability, especially for highly radioactive materials and alpha-particle emitters, is essentially unique to the national laboratories—comparable capability does not exist in universities or the private sector in this country. If the capability for handling radioactive materials in the national laboratories were lost, it would effectively halt EM-relevant R&D.

Radiochemical laboratories are typically restricted areas accessible only by qualified personnel. They have specially designed ventilation and waste handling systems. Laboratory air is constantly monitored for contamination. Personnel usually must wear protective clothing and be monitored when exiting to ensure they are free from contamination.

Containment facilities in radiochemical laboratories typically include radiochemical hoods and glove boxes. Radiochemical hoods allow workers to handle low levels of radioactive materials in essentially the same way that hazardous, nonradioactive chemicals are handled in commercial and university laboratories. Glove boxes are literally large metal boxes equipped with gloves and transparent windows. Workers insert their hands into the gloves to handle larger amounts of radioactive materials than can be handled safely in hoods (Figure 3.1). Radiochemical hoods and glove boxes are suitable for radioisotopes that do not emit penetrating radiation (e.g., Pu-239 and other primarily alpha-particle-emitting isotopes), but their limited shielding generally allows use of only tracer-level amounts of gamma-emitting isotopes.

Shielded cells, often referred to as "hot cells," allow safe handling of full levels of radionuclides that produce penetrating gamma or neutron radiation (e.g., actual tank waste, spent fuels). They feature thick concrete shielding walls, thick (typically 3 feet) multilayer leaded-glass windows, and remote manipulators (Figure 3.2). Other controls such as personnel access, monitoring, and ventilation are equal to or more rigorous than for radiochemical laboratories. The national laboratories visited by the committee have shielded cells, although there are differences in their design and potential uses. For example, the Irradiated Fuel Examination Lab (Building 3525) at ORNL can accept full-length light-water reactor fuels. It was used by EM for materials packaging from 1999 to 2003.

FIGURE 3.1 Glove box in a radiochemical laboratory. Glove boxes provide safe containment for laboratory work with radioactive or chemically hazardous materials. Substantial amounts of radionuclides that do not emit penetrating radiation (e.g., Pu-239) can be handled in a glove box. Glove boxes offer good visibility and access, but the thick, often lead-lined gloves limit the manual dexterity of scientists and technicians.
SOURCE: Department of Energy.

FIGURE 3.2 Hot cell work with remote manipulators. Heavily shielded facilities, often called hot cells or caves, allow work with full levels of radioactive materials such as high-level waste and irradiated nuclear materials. Concrete walls and leaded-glass windows are typically 3 or more feet thick to provide shielding. The windows are often filled with oil to improve visibility, which is remarkably good, and to stop some types of radiation. Skilled operators can replicate most types of hands-on laboratory work with the manipulators, but this requires much training, experience, patience, and ingenuity.
SOURCE: Department of Energy.

Support services include radiochemical laboratories to provide sample analyses, standardized and quality-controlled radioactive sources, and equipment calibration. Monitoring, dosimetry, and other worker protection services—often referred to as radiation protection or health physics—are also required. Service organizations may simply support R&D work, but more often they also perform their own R&D, for example, to improve radiochemical analyses, radiation detection, and understanding of radiation health effects. All national laboratories visited provide such support services. SRNL highlighted special capabilities for high-sensitivity measurements of ultra-low levels of radioactivity.

Maintaining Capabilities for EM

As applied to the gaps identified in Chapter 2, R&D in radiochemical laboratories and shielded cells may pertain to:

- Tank waste—its basic chemistry and rheological properties and how it can be processed;
- Radioactive contaminants in groundwater—their basic chemistry and their interactions with geologic media and microbes; and
- Radioactive contaminants remaining in facilities to be decontaminated—their basic chemistry and interactions with construction materials such as steel and concrete.

Glove boxes and shielded cells will typically be used when basic R&D or process development requires use of actual waste. Transuranic-contaminated waste might be handled in glove boxes, whereas high-level tank waste must be handled in shielded cells. Some D&D work may require glove boxes or shielded cells.

Radioanalytical laboratories house the same types of instruments used in well-equipped chemistry laboratories, such as inductively coupled plasma optical emission and mass spectroscopy, digital autoradiography, bulk and micro x-ray diffraction, scanning and transmission electron microscopy with wavelength dispersive spectroscopy, electron microprobe, liquid and ion chromatography, capillary electrophoresis, and multipoint surface area analysis. These instruments may themselves be contained in radiochemical hoods or glove boxes. Using these instruments to analyze radioactive samples usually requires that they be dedicated to this use (i.e., the instrument is considered to be contaminated with radioactive materials so it can be operated only in a radiologically controlled area). Obtaining and maintaining dedicated instruments is a substantial financial burden. Environmental samples, such as those from groundwater wells, may require analysis in low-background laboratories where low levels of radionuclides can be quantified.

Scientists and engineers of all disciplines who are engaged in EM work are likely to do at least some of their work with radioactive materials. They become qualified to work with radioactive materials through onsite training and experience, although radiochemists, for example, may have done somewhat similar work at universities.

Trained and experienced technicians and operators are essential for conducting work with radioactive materials. They understand and enforce strict procedures for handling these materials. Their skills are often unique, and like other crafts, they are learned from more experienced personnel. A year or more of experience may be necessary to become competent in operating an apparatus while wearing thick gloves in a glove box or in handling glassware with remote manipulators.

Much of the infrastructure to handle radioactive material at DOE sites (e.g., the Radiochemical Processing Laboratory at PNNL, the Radiochemical Engineering Development Center at ORNL) is decades old. These facilities degrade over time without adequate maintenance and programs to utilize them. In some cases some key facilities (e.g., the Radiochemical Processing Laboratory) have been threatened with shutdown. In recent years support for such facilities has improved. However, despite EM's need for them, agencies other than EM, such as DOE's Office of Science (SC) and National Nuclear Security Administration and the Department of Homeland Security, are providing most of their support (PNNL 2007).

CONDUCTING ENGINEERING AND PILOT-SCALE TESTS

EM's engineering and technology development includes testing to provide basic parameters (e.g., heat and mass transfer, mixing, corrosion) to design new processes and equipment and to demonstrate them at the pilot scale or larger.[1] Capabilities for engineering and pilot-scale testing are needed to support R&D to address all of the gaps identified in Chapter 2, especially WP-1 through WP-5, GS-2, GS-4, DD-1, and DD-3.

Engineering test facilities, sometimes referred to as semiworks, are used throughout the private sector, and all the national laboratories the committee visited have them. EM contractors and universities often use their own facilities for EM projects. As one example, Clemson University tested prototype glass melters for the Savannah River Sight (SRS) Defense Waste Processing Facility (DWPF). Engineering facilities generally do not allow the use of radioactive materials, although two important exceptions are described in this section.

Tank mock-up facilities are onsite capabilities that EM will need for many years to support waste tank cleaning. Retrieval of tank waste is a major challenge at Hanford and SRS. Both sites have full-diameter tank mock-ups used by contractors and national laboratory personnel to test retrieval technologies (e.g., pumps, high-pressure water lances, robotic devices). The mock-ups allow the simulation of limited equipment access to the tank interior, as is the case for the actual tanks, and for SRS, the ability to reproduce the complicated internal cooling coil geometries that make tank cleaning especially challenging for that site (Chapter 2, Figure 2.3). INL used a mocked-up tank floor for testing grout flow and emplacement methods to encapsulate sludge heels.

---

[1] Pilot scale typically refers to testing with kilogram quantities of materials or more, up to perhaps half of the production of the full-scale process.

High-bay buildings, large buildings with sections that provide several stories of overhead space, are necessary for onsite equipment fabrication and testing. Much of the equipment used in EM cleanup work is physically large, especially waste processing equipment. PNNL is testing pulse jet mixers, special devices for mixing liquids and sludges in Hanford's Waste Treatment Plant, in its high-bay building 336 (Figure 3.3). SRNL highlighted its engineering development laboratory during the committee's site visit. SRS also operates a mock-up facility where every component to be installed in the DWPF is pretested.

Engineering laboratories are necessary for testing materials and equipment components to ensure process safety, operability, and reliability. Examples of materials and component tests include fatigue or fracture under high-temperature, -pressure, -stress, or corrosive conditions. Small-scale versions of new equipment or processes are often set up and tested in engineering laboratories. Tests for quality control and quality assurance are also included. Such capability is common in the private sector. Onsite capability

FIGURE 3.3 High-bay building for engineering tests. High-bay buildings provide two or more stories of vertical space for testing or demonstration of large equipment, which is typically required for waste processing. In this photo, pulse jet mixers are being tested for use in Hanford's Waste Treatment Plant. Thorough testing of such newly designed equipment is necessary because it must operate reliably as designed, and with little or no opportunity for maintenance, once placed into radioactive service.
SOURCE: Department of Energy.

for EM work is needed as a practical matter, and all national laboratories visited have engineering laboratories.

Radioactive semiworks for pilot-scale testing of large-scale processes are highly desirable but usually infeasible due to construction time and cost. Notable exceptions are the actinide removal process (ARP) and the modular caustic-side solvent extraction unit (MCU) that are being operated at SRS until its Salt Waste Processing Facility (SWPF) is completed in about 2013. Processing the salt portion of SRS tank waste has been delayed for a variety of reasons. The ARP and MCU were built primarily to process some of the salt because the site's waste tanks are almost full (Appendix G; NRC 2006b). These facilities can provide data and operating experience to help ensure that the SWPF meets its performance objectives—a major risk reduction opportunity for EM noted in gap WP-2.

Maintaining Capabilities for EM R&D

Engineers of essentially all disciplines, along with technicians and operators, typically use and maintain engineering-test and pilot-scale facilities for their experimental work and process demonstrations. Mock-ups may test equipment or processes at pilot scale or full scale. Since high-bay buildings already exist on sites and at national laboratories, keeping them, rather than demolishing and rebuilding as programs change, is probably cost-effective.

Assembling equipment for engineering tests and operating it successfully require a good deal of experience and technical savvy among technicians and operators. The capabilities to accurately machine special alloys, weld them, and operate high-pressure devices are examples. Accumulating this knowledge may take years of hands-on experience and mentoring from more experienced personnel. Further, if the technicians and operators have experience with the site problems that their project is addressing, they often contribute innovative, practical ideas toward their solutions.

DETERMINING CONTAMINANT BEHAVIOR IN THE ENVIRONMENT

Each of the DOE sites has a unique history in the disposal or release of contamination and unique geohydrological characteristics, which largely control the movement of these contaminants. Contamination has reached the groundwater at all four sites visited. Groundwater and soil remediation are in progress and will continue for the duration of the EM cleanup. Capabilities for determining contaminant behavior in the environment are needed to support R&D to address the groundwater and soil (GS) gaps identified in Chapter 2.

These capabilities are widely available in the private sector and universities. EM contractors and national laboratory and university researchers are often partners in projects aimed at understanding contaminant behavior at the DOE sites and in conducting remediation projects. Sampling, monitoring, and implementing remedial actions must, of course, be done on the site itself.

Field test facilities that provide actual data on contaminant behavior in the environment, often referred to as "contaminant fate and transport," are an essential and unique capability for the sites and national laboratories. The needs are specific to each site due to the individual site histories and discharges of contaminants, types of contaminants, site characteristics, and possibilities of future releases from storage or disposal facilities (e.g., waste tanks, capped trenches). Field test facilities can include physical structures in designated areas of the site, or they may simply be monitoring wells or stream sampling points located on- and off-site.

All national laboratories visited have field test facilities and are actively conducting field tests to determine contaminant fate and transport at their associated sites. ORNL highlighted its Field Research Center during the committee's site visit (Figure 3.4). SRNL described site monitoring and

FIGURE 3.4 Oak Ridge Field Research Center. Site-specific data are required for characterizing geohydrology and measuring contaminant transport and effects of remedial actions. A specific location on a site, including dedicated boreholes, water wells, and equipment, may be developed for this purpose. Monitoring wells located around the site and surface water sampling also provide field data. Geohydrological parameters usually change slowly, so maintaining these facilities for years or decades is necessary.
SOURCE: Department of Energy.

field tests of various remediation technologies, including bioremediation (Appendix G). INL described enhanced bioremediation tests, which involve injecting microbes and nutrients into a carbon tetrachloride plume source ("hot spot") and monitoring the groundwater (Appendix E).

PNNL is operating the Hanford 300 Area Integrated Field Research Center, which is funded by SC and is intended to provide a fundamental understanding of coupled geochemical, hydrologic, and microbiologic processes in the contaminated aquifer that will enable development of an effective, long-term remedial strategy for uranium at the site.[2] PNNL is also testing an engineered barrier ("Hanford cap") to provide long-term control of contaminant migration from buried waste (Appendix D).

Information archives that maintain the long-term accumulated knowledge relevant to understanding contaminant fate and transport at the cleanup sites are unique capabilities of the national laboratories. Each site has an essentially permanent relationship with a colocated national laboratory, dating back to the establishment of the site. Data on waste disposals, contaminant releases, and environmental monitoring have accumulated over the years and will continue to do so. Along with this is the growing understanding of the site characteristics that govern fate and transport. Such information comes from site contractors and university research as well as from the national laboratories. However, only the national laboratories have the long-term capabilities to maintain and synthesize all of this information into sufficiently detailed conceptual understanding and site models to guide EM's remediation work and DOE's long-term stewardship planning.

Geoscience and geotechnical laboratories that support site cleanup are often equipped for handling low levels of radionuclides as well as for engineering tests, for example, tests to determine sorption of contaminants onto soils and rocks and their permeabilities. Geotechnical laboratories typically are part of the national laboratory infrastructure for handling radioactive materials and conducting engineering tests described in the previous two sections. INL highlighted its geocentrifuge, which allows accelerated tests of flow through geologic media.

Maintaining Capabilities for EM

Site and national laboratory facilities for environmental studies typically are shared freely among national laboratory, university, and other researchers engaged in this work. Environmental scientists, geoscientists, chemists, and engineers are typically involved in contaminant fate and transport studies.

---

[2] See http://ifchanford.pnl.gov/.

Field test facilities at the sites are unique in the sense that they cannot be replicated elsewhere to measure the same phenomena. Field tests, for example, of the engineered barriers described in GS-3, must usually be run for years before they provide useful information. The Nuclear Regulatory Commission suggested exhuming an SRS saltstone lysimeter, which operated over 20 years ago, to help resolve some questions about saltstone performance; see gap GS-4.

Professional researchers and technicians require years to become fully acquainted with the geohydrological characteristics of a site and how they have affected the fate and transport of contaminants released to the site. Information archives would include not just physical databases, but also experienced personnel to interpret and build on accumulated knowledge. For a 30-year program with many experienced personnel now retiring, accumulated site knowledge will have to be passed on through perhaps two generations of new scientists.

UTILIZING STATE-OF-THE-ART SCIENCE TO DEVELOP ADVANCED TECHNOLOGIES

The national laboratories maintain extensive and diverse world-class scientific capabilities that are supported primarily by the DOE SC. Presentations by the national laboratories during the committee's site visits and by SC during the committee's April 2008 meeting (Appendix B) provided an overview of these capabilities. Clearly SC capabilities are necessary for EM's engineering and technology development. In addition, it is clear that the state of the art will advance over the next 30 years of the EM cleanup in ways that cannot be imagined today. While EM would not be expected to be a primary user or financial supporter of advanced scientific facilities, EM's and SC's continued close cooperation and coordination can ensure that EM is able to utilize state-of-the-art science (Chapter 4). As a DOE office, SC shares with EM the responsibility for protecting citizens and the environment from deleterious effects of DOE's legacy of nuclear materials production.

A particularly important SC-funded resource for EM-related studies has been the capability for x-ray and infrared spectroscopies, microspectroscopies, and tomography available at the nation's synchrotron light sources (including the Advanced Photon Source, the National Synchrotron Light Source, the Advanced Light Source, and other sources located at DOE laboratories).[3] Through studies led by researchers from universities

[3] EM researchers have made significant use of DOE SC synchrotron facilities and the Environmental Molecular Science Laboratory (EMSL) to determine the spatial locations, mineral associations, and chemical nature of DOE contaminants in subsurface sediments from vadose zone and groundwater plumes. The resulting scientific information has been im-

as well as from federal laboratories, these capabilities and infrastructure have identified the chemical environment of contaminants on soil minerals using spatially resolved x-ray fluorescence (XRF) and x-ray absorption (XAS) spectroscopies; identified the speciation of Pu, U, and other heavy metals in soils and sediments, in association with plants and microbes, and on engineered surfaces requiring decontamination; and aided in the evaluation of chemical remediation strategies through providing data to better model the mobility and fate of contaminants. For example, Los Alamos researchers were able to use x-ray absorption near edge spectroscopy to identify the speciation of plutonium in contaminated soil and concrete samples from the Rocky Flats site, data that were very valuable in informing cleanup efforts (LANL 2002). Another recent study cited the use of XAS at the Advanced Photon Source at Argonne National Laboratory to identify the chemical and mineral state (critical to understanding mobility) of uranium beneath high-level waste tanks at the Hanford site (Catalano et al. 2006). Microprobe XRF measurements have been conducted at the National Synchrotron Light Source at Brookhaven National Laboratory on treated sediment samples from the Savannah River and the Hanford sites to study the effect of phosphate and microbes on removal of uranium to develop improved technologies for remediation (Knox et al. 2008).

Development of waste separations technology has also benefited from synchrotron analysis: XAS studies at the Stanford Synchrotron Radiation Laboratory have been used to characterize the composition of Np- and Pu-containing waste-sludge alkaline-wash solutions, identifying highly soluble species and leading to design of enhanced chemical separations processes (Neu et al. 1999). Overall, by allowing university and national laboratory researchers to both thoroughly investigate samples taken from DOE sites and create controlled experiments with model matrices, surfaces, and sets of conditions analogous to in situ environments, synchrotron and complementary surface and molecular spectroscopies are extremely valuable resources for the DOE's EM mission.

Advanced computing is an overarching capability to support all facets of EM engineering and technology development. Such capability includes modeling waste inventories and interactions, treatment processes and process design, site geohydrology, and contaminant fate and transport. Advanced computing is a basic capability required for conducting state-of-the-art science, and all national laboratories visited have powerful computers for basic and applied research. Those at ORNL and PNNL were

---

portant in the definition of the geochemical state of sorbed contaminants and is the first step in devising a remedial strategy. Use of these state-of-the-art facilities for EM's radioactively contaminated samples, however, required the facilities to implement expensive safety and health procedures.

highlighted during the committee's site visits. Computing capabilities are essential to address all of the gaps described in Chapter 2, especially WP-2, WP-3, GS-1, GS-2, and DD-3.

Surface analyses are another example of state-of-the-art capabilities that can help EM address its engineering and technology gaps. Surface analyses are important for understanding the chemical and physical interactions of contaminants with the primary materials of interest for D&D projects (concrete, stainless steel, paints, strippable coatings), waste form development (glasses, ceramics), and environmental studies (soils, biofilms) to gain a better understanding of how contaminants bind to and penetrate these materials. Some concrete and steel surfaces in DOE structures have been in contact with radioactive materials for 60 years.

Surface analytical capabilities include those at the EMSL at PNNL such as time-of-flight secondary ion mass spectroscopy, infrared spectroscopies, and high-sensitivity surface probe microspectroscopies. SRNL reported surface analytical capabilities, including glove-box-contained electron microscopies and vibrational and electron spectroscopies. ORNL highlighted its spallation neutron source and high flux isotope reactor for materials studies. In many cases, custom-designed systems with high sensitivities and intensities not normally available in commercial instruments have been developed through significant expenditures of effort on the part of DOE-supported researchers. Surface analysis and spectroscopic capabilities are particularly important to waste form development, WP-5; understanding long-term behavior of cementitious materials, GS-4; and removing contamination from surfaces, DD-3.

CONCLUSIONS

The capabilities (personnel expertise and physical infrastructure) described in this chapter are those that the committee judged to be necessary to support R&D to address the science and technology gaps identified in Chapter 2. The committee intentionally highlighted those capabilities that are essentially unique to the DOE sites and national laboratories. Most are important resources for other DOE programs as well as those of EM, and many are in fact being utilized by other DOE offices such as the Office of Nuclear Energy. Partnering with these other programs to leverage EM's engineering and technology initiatives is described in Chapter 4.

4

Leveraging R&D for Environmental Management

The statement of task for this study asks the committee to identify strategic opportunities to leverage research and development (R&D) from other Department of Energy (DOE) programs, other federal agencies (e.g., Department of Defense [DOD], Environmental Protection Agency [EPA]), universities, and the private sector.[1] "Strategic" in this statement is interpreted by the committee in the same sense that "strategic" initiatives are set forth in the Office of Environmental Management's (EM's) Engineering and Technology Roadmap (DOE 2008b). In the Roadmap, strategic initiatives are those that address the technological risks and uncertainties identified by EM.[2] The strategic opportunities for leveraging discussed in this chapter would help EM bring R&D from other organizations to bear on the technology gaps identified by the committee in Chapter 2.

For purposes of this report "opportunities to leverage" are defined as opportunities for collaborations or co-investments between EM and other organizations—government, academic and private sector—to achieve synergistic production of new knowledge, knowledge transfer and application to cleanup problems, reduction in time schedules, and efficiency improvements in personnel and infrastructure utilization. Such synergy requires that the participants achieve tangible benefits and outcomes, which may include

[1] See Chapter 1, Sidebar 1.2.

[2] The EM Engineering and Technology Roadmap will be referred to as the EM roadmap, or simply as the Roadmap. The committee's statement of task used the term "technology gaps," which are discussed in Chapter 2, rather than "technical risk or uncertainty," which is used in the Roadmap.

reduced costs, accelerated R&D and cleanup schedules, improved technology transfer, workforce maintenance, and facility support. Through such collaborations EM can leverage its R&D investments (financial, personnel, and management commitment) to strengthen its partners' R&D programs as well as to improve its own site cleanup work (Sidebar 4.1).

When identifying leveraging opportunities the committee focused its information gathering on programs at potentially relevant federal agencies. The reason for doing so is that federal agencies fund virtually all R&D relevant to EM at the national laboratories and universities. As a consequence, focusing on federal programs was an efficient, nonduplicative way to identify relevant leveraging opportunities. Pursuing leveraging opportunities might begin by contacting managers of various federal R&D programs but would then lead to contact with individuals in the organizations actually performing the R&D.

SIDEBAR 4.1
What Is Leveraging?

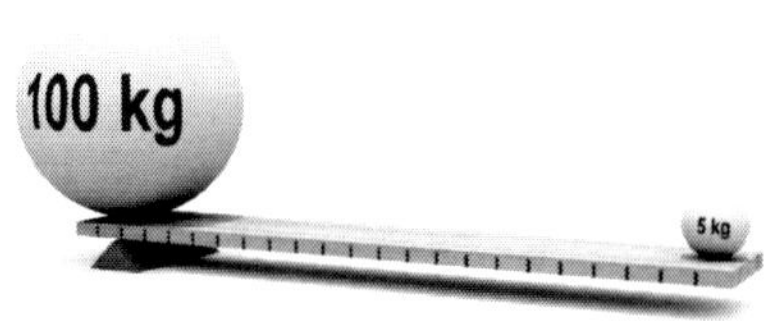

Leveraging, a word in common usage, describes all sorts of activities in which resources used are magnified in the outcome. "Leveraging" is derived from the word "lever," which is a simple device that provides mechanical advantage through the use of a fulcrum. A small force at a great distance from the fulcrum can be magnified to balance a large force over a small distance on the other side of the fulcrum.

The pooling of many small efforts into an effort that addresses a common issue leverages each small effort by a multiplier that is the sum of the number of small efforts. Thus five small coordinated efforts pooled leads to a leverage of five times for any one of the small contributors—provided the pooled effort addresses their specific need. Conversely, a large (central) effort or capability can be a source of support for a number of smaller satellite uses of this capability, where any single satellite could not support the capability it needed to complete its work. In practice, there are many variants of this principle that require some form of "partnering" or collaboration among the component organizations.

The research management literature contains many descriptions of leveraging experiences between organizations in the private, governmental, and academic sectors with lessons learned from such experiences. For example, a search of the past 10 years in the journal *Research-Technology Management,* a publication of the Industrial Research Institute,[3] generated over 175 references to "leveraging" R&D activities with descriptions of both the value and the shortcomings of such activities. A review of these articles and discussions with Michael Dalton and Charles Scouten, consultants in the field of technology management, at the committee's April 2008 meeting (Appendix B) brought out the fact that the more successful of these activities have characteristics that include a strong, sustained commitment, a disciplined approach, and a detailed process from research to implementation by the partners to ensure success of the collaboration.

Examples of effective and successful leveraging involving Los Alamos National Laboratory (LANL) and the private sector are summarized in Sidebar 4.2. Such leveraging partnerships benefited the partners in different ways. In this example, the partnerships provided LANL with a mechanism for stabilizing its R&D funding, a test bed for some of its analytical capabilities, and the stimulation of applying its research capabilities to problem areas outside the nuclear complex. For the private-sector companies involved, leveraging provided access to multidisciplinary expertise to address their complex problems as well as pathways to solve these problems.

Reviewing EM's past experience with leveraging partnerships, Gerald Boyd, manager of the DOE Oak Ridge Operations Office and a former manager of EM's Office of Science and Technology, discussed his earlier experiences with EM technology development and summarized guiding principles for success. His summary of guidelines for success (Sidebar 4.3) reflected many of the same factors described above and in studies of successful partnerships in many organizations as described by Slowinski and Sagal (2003).

The Environmental Management Science Program (EMSP) was a unique program that leveraged research capabilities between the national laboratories and universities to address problems defined by EM. It was established in 1995 and funded research until 2003 when the program was transferred to the Office of Science (SC) to become part of the Environmental Remediation Sciences Division (ERSD), described in the next section of this report. At the height of the EMSP initiative, EM defined the research challenges through focus area teams at each site. These research challenges were used to define requests for proposals that were issued to the national laboratories, universities, and industry. Proposals that were multi-institutional were encouraged, especially for national laboratories

[3] See http://www.iriinc.org/.

SIDEBAR 4.2
Los Alamos National Laboratory Industry Alliances

The partnership alliances between LANL and industry were the result of a program established at LANL several years ago to manage its intellectual property in order to (1) partner with industry to enhance science in the service of national security, (2) strengthen the U.S. economy by accelerating product creation from LANL technologies, and (3) foster technology job growth in the New Mexico regional economy. Partners were chosen who had characteristics that matched the needs of the laboratory both in technology development and market coverage. The objective was to establish a long-term relationship in which the technological and financial needs of both organizations could be met. The process of cultivating an effective partnership took substantial effort on the part of both organizations in order to align time horizons, develop shared values, agree on investments to be made by both parties, identify and use best practices in partnering, and make a long-term commitment to invest in developing the relationships necessary to succeed.

Two such partnerships, one with Chevron and another with Proctor and Gamble, are examples in which the expertise of each organization is leveraged through the other partner. In the case of Chevron, one example of technology leveraging was the successful utilization of LANL technology for secure battlefield communication adapted to oil-field monitoring. In another project under the partnership, improved drilling fluids were developed in collaboration with Baker Hughes and Lucite companies based on LANL's analytical capabilities in understanding high-pressure fluid dynamics. The alliance with Proctor and Gamble, on the other hand, applied reliability software developed as part of the weapons program to predicting reliability in a complex consumer manufacturing process, resulting in capital savings of over $2 billion.

In both partnerships, the participants recognized the importance of having an effective process for partnering and adopted a set of common operational principles. Such alliance principles are described in literature on management, most recently in a book entitled, *The Strongest Link: Forging a Profitable and Enduring Corporate Alliance* (Slowinski and Sagal 2003).

SOURCE: Freese (2008). LANL, April 28, 2008, presentation to this committee.

that partnered with universities and private companies. Once selected for funding, the principal investigators were required to meet periodically with staff from the EM focus areas to listen to current problems associated with their technical area (e.g., tank waste chemistry, groundwater remediation, decontamination and decommissioning) and to describe the progress that had been made on the funded research project. This approach allowed for the development of fundamental research that was focused on specific EM problems, and engaged the national laboratories, industry, and colleges and universities.

SIDEBAR 4.3
Guiding Principles for Successful Partnerships in Previous EM Technology Development[a]

Principles for successful partnerships learned from the EM focus areas and the EM Science Program[b] were the following:

• Communication among developers, end users, regulators, and stakeholders: Specifically, scientists must take responsibility for problem resolution; engineers, in turn, must realize when information is not sufficient to allow for a defensible remedy.
• Early identification of technology and technology needs during project planning to allow funding and schedule allowances;
• End-user input/involvement in design, development, and testing of new technologies;
• Integration of field technology team and field project management team; and
• Teaming between technology developers and engineering companies performing the field work.

[a]Boyd (2008).
[b]NRC (1997a, 1999c).

LEVERAGING OPPORTUNITIES FOR EM TO ADDRESS TECHNOLOGY GAPS

At its April 2008 meeting, the committee received input from organizations with which EM has worked or potentially could collaborate in addressing the technology gaps that were identified in Chapter 2. The organizations that participated in discussions and provided input to the committee are listed in Sidebar 4.4. Their capabilities relevant to EM's needs are summarized in this section. These leveraging opportunities are given as examples and are not intended to be a comprehensive list.

Offices Within the Department of Energy

Office of Science

SC is the single largest supporter of basic research in the physical sciences in the United States, providing more than 40 percent of the total funding. SC manages fundamental research programs in basic energy sciences, biological and environmental sciences, and computational science. In addi-

SIDEBAR 4.4
Organizations That Participated in the Committee's Meeting on Opportunities for EM to Leverage Its R&D Programs

Department of Energy[a]

- Office of Science
- Office of Civilian Radioactive Waste Management
- Office of Nuclear Energy

Other federal organizations

- Department of Defense
- Department of Homeland Security
- Environmental Protection Agency
- Nuclear Regulatory Commission

International organizations

- International Atomic Energy Agency
- Nuclear Energy Agency of the Organisation for Economic Co-operation and Development

[a]Three DOE national laboratories: Oak Ridge, Pacific Northwest, and Savannah River also made presentations.

tion, SC is the federal government's largest single funder of materials and chemical sciences, and it supports unique and vital parts of U.S. research in climate change, geophysics, genomics, life sciences, and science education.[4] With a budget of roughly $4 billion dollars in 2008, approximately one-third of that amount was invested in colleges and universities in the form of research grants, with over 300 institutions of higher education receiving such awards.[5]

DOE's laboratories and technology centers that receive their primary funding from SC house world-class facilities where more than 30,000 scientists and engineers perform cutting-edge research.[6] The national laboratories, including the four associated with the DOE sites that are part of this study, are important sources of expertise and technology as they relate

[4] See http://www.er.doe.gov/about/index.htm.

[5] See http://www.science.doe.gov/SC-2/Presentations/Blevins%20NCURA%20Nov%203%202008.ppt#380,10,Office of Science Numbers.

[6] See http://www.doe.gov/organization/labs-techcenters.htm.

to the nuclear industry. In particular, the four national laboratories have years of experience as well as unique capabilities to address the technology gaps in EM's program (see Chapters 2 and 3).

At the April 2008 meeting, Mike Kuperberg (2008) stated that the SC mission is to deliver the remarkable discoveries and scientific tools that transform our understanding of energy and matter and advance the national, economic, and energy security of the United States.[7] Strategic Goal 2 within this mission is to provide the biological and environmental discoveries necessary to clean and protect our environment, offer new energy alternatives, and fundamentally alter the future of medical care and human health. Strategy 2.3 in the SC strategic plan is to "understand the complex physical, chemical and biological properties of contaminated sites for new solutions to environmental remediation."[8] The plan details a number of the unresolved issues stemming from the legacy wastes at the DOE sites.

Kuperberg (2008) described a generic model for SC collaboration with DOE's program offices such as EM as a research "continuum." The model begins with discovery research and use-inspired basic research, which are mainly the purview of SC. The model then moves to applied research and, lastly, technology maturation and deployment, which are mainly the purview of the DOE technology program offices. This continuum is by no means linear, recognizing that technological challenges can identify opportunities for basic research and breakthroughs in basic research can accelerate advances in new technologies.

Kuperberg also described offices within SC and their missions that are relevant to EM technology development:

- The Office of Basic Energy Sciences (OBES) has the mission to foster and support fundamental research to expand the scientific foundations for new and improved energy technologies and for understanding and mitigating the environmental impacts of energy use. The OBES research portfolio includes material sciences and engineering; chemical and geo- and biosciences; and scientific user facilities, for example, neutron scattering facilities.
- The Advanced Scientific Computing Research program has the primary mission to discover, develop, and deploy the computational and networking tools that enable researchers in the scientific disciplines to analyze, model, simulate, and predict complex phenomena important to DOE.
- The Office of Biological and Environmental Research (OBER) has

[7] Also see http://www.er.doe.gov/about/Mission_Strategic.htm.

[8] DOE Office of Science Strategic Plan 2004. See http://www.er.doe.gov/about/Strategic_Plan/Feb-2004-Strat-Plan-screen-res.pdf.

the mission to advance environmental and biological knowledge that promotes national security through improved energy production, development, and use. ERSD was established within OBER in 2003 to provide fundamental science to support DOE's long-term cleanup challenges.[9] Of OBER's operating budget for research ($45 million for fiscal year 2008), approximately 28 percent was invested in colleges and universities through research grants primarily within the ERSD.

As examples of engagement with EM problems, Kuperberg (2008) described OBES work in developing the "BOB Calix" extractant, which was a major breakthrough for the Savannah River Site (SRS) salt waste treatment; studies of the hydrolysis of plutonium; and effects of ionizing radiation on uranium (IV) peroxides. SC's computing capability is being applied to models of uranium transport within the Hanford 300 Area and to hybrid models that describe contaminant transport.

OBER's Environmental Remediation Science Program (ERSP), which consolidated OBER's Natural and Accelerated Bioremediation Research and the former EMSP, focuses on DOE-relevant contamination, understanding fate and transport, new remediation concepts, and monitoring. The committee noted that much information has been developed by OBER-supported fundamental, mission-orientated investigations into the fate, transport, and remediation of metals and radionuclides. This information and that developed by ongoing and planned ERSP research is relevant to EM needs. EM involvement in the review process of new and ongoing ERSP projects could facilitate the more rapid translation and implementation to achieve cleanup.

Another opportunity for EM's leveraging noted by the committee is the analysis of the results of past and current research outcomes from OBER-sponsored projects. These results could be systematically examined to identify those niche- and case-specific parameters that operationally affect successful implementation of new cleanup technology. Often the scientific and technical nuances of fundamental research findings may be marginalized in attempting to scale up or implement a new technology at a new site or location, particularly when transferred to a cleanup contractor. The use of a technology assessment teams consisting of the contractor, EM staff, and OBER investigators could help ensure that fundamental information is most effectively implemented or that EM identifies those gaps that must filled to increase the probability of successful implementation of new and innovative technology.

Kuperberg (2008) noted that there is close cooperation between ERSD

[9] In early 2009, ERSD was consolidated with the Climate Change Research Division to create the Climate and Environmental Sciences Division.

and EM's Office of Engineering and Technology (EM-20). To more effectively leverage their R&D, it will be important to integrate the EM roadmap and SC's initiatives, for example, the Strategic Timeline for Biological and Environmental Research contained in the SC Strategic Plan. EM can be a full partner in defining the programs by which SC will fulfill its Strategy 2.3, which bears directly on EM's responsibilities. Leveraging its R&D programs with SC offers EM an important vehicle for further partnering with universities and the private sector.

Office of Fuel Cycle Management of the Office of Nuclear Energy

The mission of the Office of Nuclear Energy (NE), as described to the committee by Andrew Griffith, Acting Director, Recycled Fuel Development (NE-53) is to lead the DOE investment in the development and exploration of advanced nuclear science and technology (Griffith 2008). NE leads the government's efforts to:

- Develop new nuclear energy generation technologies,
- Develop advanced, proliferation-resistant nuclear fuel technologies that maximize energy from nuclear fuel, and
- Maintain and enhance the national nuclear technology infrastructure.

NE aims to serve the present and future energy needs of the nation by managing the safe operation and maintenance of the DOE nuclear infrastructure. NE manages the Global Nuclear Energy Partnership (GNEP) program, which is intended by the United States as a cooperation with other nations to develop and deploy advanced nuclear recycling and reactor technologies. At the committee's April 2008 meeting, NE presented GNEP as a major component of its fuel cycle technologies program.

Overall, NE's advanced fuel cycle initiatives draw on experience from across the DOE complex, including at least 10 national laboratories (Griffith 2008). Facilities for NE work at national laboratories include those for handling radioactive material and for conducting engineering tests, which were described in Chapter 3 as national laboratory resources for EM work. These clear overlaps between NE's needs for expertise and infrastructure and those of EM suggest that cooperation in maintaining these capabilities is essential to both offices.

Office of Civilian Radioactive Waste Management

The mission of DOE's Office of Civilian Radioactive Waste Management (RW) is to manage the nation's high-level radioactive waste and spent nuclear fuel. Its focus is the licensing and development of the proposed Yucca Mountain geologic repository in Nevada. Jeffrey Walker, RW, gave an overview of the status of the repository and described several opportunities for cooperation between EM and RW. Walker (2008) stated that the scientific basis for Yucca Mountain's licensing was complete. Opportunities for cooperation include:

- Waste package technology,
- Waste handling, and
- Performance monitoring and confirmation.

Waste package technology could be improved by developing less costly corrosion-resistant materials and more efficient manufacturing and testing methods. Waste must be handled remotely, which suggests needs for improved robotic technologies. Innovative sensor technology and remote monitoring capabilities will be important for confirming the repository's performance before and after closure. Ensuring worker safety in both the aboveground and subsurface waste handling operations is central to RW's program.

Other Federal Organizations

Strategic Environmental Research and Development Program of the Department of Defense

The Strategic Environmental Research and Development Program (SERDP) is the DOD environmental science and technology program, planned and executed in full partnership with DOE and EPA, with participation by numerous other federal and nonfederal organizations. SERDP has environmental drivers that call for the reduction of current and future environmental liabilities.

In his presentation, Bradley Smith, SERDP executive director, highlighted environmental challenges including the current intractability of remediating some chlorinated solvents, DOD's potential liability for unexploded ordnance, and emerging new contaminants such as perchlorates. Smith described SERDP as supporting R&D up through proof of principle. Linked with SERDP in a combined program office is the Environmental Security Technology Certification Program (ESTCP), which moves promising new environmental technologies to the demonstration phase and promotes their implementation. Smith (2008) also described a roadmap that had been used previously by SERDP and ESTCP for environmental restoration of DOD sites.

Environmental restoration needs of DOD are broadly similar to those of EM, which were described in Chapter 2. DOE is a partner in planning and carrying out SERDP, as noted above. The relation between SERDP and ESTCP appears conceptually to mirror that between SC and EM. ESTCP functions in a similar way as EM's previous focus areas, which were described by Boyd (2008).

Domestic Nuclear Detection Office of the Department of Homeland Security

William Hagan, assistant director of the Domestic Nuclear Detection Office (DNDO), described his agency's functions as a national office established to improve the U.S. capability to detect and report unauthorized attempts to import, possess, store, develop, or transport nuclear or radiological material for use against the nation, and to further enhance this capability over time (Hagan 2008). DNDO's transformational R&D program includes:

- Exploratory Research,
- Advanced Technology Demonstrations,
- Small Business Innovation Research, And
- Academic research.

Hagan stated that DNDO and EM have some overlapping research needs in the area of nuclear detection. This could potentially provide improved technology for both real-time remote assessment of radiation sources and for long-term site monitoring. Hagan noted that both DNDO and EM have needs to detect radionuclides over large areas. He also noted difficulties in assaying nuclear materials because of their being either shielded or inaccessible. As examples of new DNDO technologies of potential relevance to EM, he described technologies that provide high sensitivities, ability to detect radionuclides at a distance (large standoff), and improved algorithms to detect masked radionuclide signatures. He also described exploratory research to identify new materials for radiation detection.

Office of Research and Development of the EPA

The Office of Research and Development (ORD) provides a scientific foundation to support the EPA's mission to protect human health and safeguard the national environment. To do this, ORD:

- Performs R&D to identify, understand, and solve current and future environmental problems;
- Provides responsive technical support to EPA's mission;

- Integrates the work of ORD's science partners (other agencies, nations, private-sector organizations, academia, and international organizations); and
- Provides leadership in addressing emerging environmental issues and in advancing the science and technology of risk assessment and risk management.

Randall Wentsel, national program director of ORD's Land Research Program, described several groundwater research areas that are relevant to EM cleanup, including:

- Dense nonaqueous phase liquid (DNAPL) source zone remediation,
- In situ treatment of source areas by thermal and chemical removal or destruction of DNAPLs and performance monitoring,
- Monitored natural attenuation for organic and inorganic contaminants, and
- Permeable reactive barriers.

These closely parallel areas suggested in Chapter 2 for EM groundwater and soil R&D.

Wentsel (2008) also stated that programs in these areas already include cooperation among federal agencies, including DOE. For example, the Inter-Agency Steering Committee on Multimedia Environmental Modeling initiated in 2001 includes six federal agencies. He listed groundwater remediation technology development, fate and transport modeling, and site characterization as areas for more joint EPA–EM R&D initiatives.

Division of Waste Management and Environmental Protection of the Nuclear Regulatory Commission

Broadly speaking, the Nuclear Regulatory Commission (USNRC) regulates the civilian use of by-product, source, and special nuclear materials to ensure adequate protection of public health and safety, to promote the common defense and security, and to protect the environment. The USNRC's regulatory mission covers four main areas:

- Reactors—Commercial reactors for generating electric power and research and test reactors;
- Materials—Uses of nuclear materials in medical, industrial, and academic settings and facilities that produce nuclear fuel;
- Waste—Transportation, storage, and disposal of nuclear materials and waste, and decommissioning of nuclear facilities; and
- Nuclear Security—Physical security of nuclear facilities and materials.

Relative to the third area within the USNRC mission, Andrea Kock, Chief of the Performance Assessment Branch, Division of the Waste Management and Environmental Protection, within USNRC's Office of Federal and State Materials and Environmental Management Programs, discussed interfaces between EM and USNRC.

As described in Chapter 2, the Ronald Reagan National Defense Authorization Act of 2005, Section 3116, authorizes DOE to declare some tank wastes to be "incidental to reprocessing," which allows those wastes to be disposed at a DOE site[10] rather than requiring disposal in a high-level waste repository (e.g., Yucca Mountain if licensed and constructed). While USNRC does not directly regulate DOE, the Reagan Act requires the DOE to consult with the USNRC on DOE's determination that criteria for an incidental waste declaration have been met. The USNRC is required to monitor DOE disposal actions. In addition to the tank heels described in Chapter 2, the saltstone disposals at SRS are subject to USNRC evaluation. Kock (2008) listed evaluation criteria for saltstone, and stated that some are technical challenges to EM.

Kock (2008) described opportunities for joint EM–USNRC long-term R&D on:

- Cementitious materials, including the ongoing DOE cement consortium;
- Ground covers, including longevity of clay covers and optimizing barrier performance; and
- More efficient approaches to cleanup problems, including optimization of groundwater models and simulation of complex source terms.

She also highlighted knowledge management initiatives, including the need to attract new technical staff and researchers to enable transfer of knowledge from retiring production-era personnel. USNRC is establishing knowledge centers in performance assessment and research to facilitate knowledge transfer and management.

International Organizations

International Atomic Energy Agency

The International Atomic Energy Agency (IAEA) was established as the world's "Atoms for Peace" organization within the United Nations in 1957. The agency works with its Member States[11] and multiple partners

[10] The Reagan Act applies only to the Idaho and Savannah River sites.

[11] IAEA Member States are sovereign nations that have formally applied and been accepted for membership in the IAEA. The IAEA had 145 Member States as of September 2008.

worldwide to promote safe, secure, and peaceful use of nuclear technologies. In pursuing its mission to promote peaceful uses of atomic energy, the IAEA's work includes:

- Promotion of research, development, and practical applications;
- Exchange of scientific and technical information;
- Exchange and training of scientists and experts;
- Establishment and administration of safeguards;
- Establishment of facilities, plants, and equipment; and
- Development of safety standards and provision for their application.

Horst Monken-Fernandes (2008) of the IAEA Division of Nuclear Fuel Cycle and Waste Technology described joint work with the IAEA Division of Radiation, Transport, and Nuclear Safety to assist IAEA Member States in four areas:

1. Development of an international safety regime for radioactive waste management,
2. Management and disposal of all types of radioactive waste,
3. Assessment and control of radioactive discharges to the environment, and
4. Decommissioning of installations and remediation of sites.

The IAEA does not carry out experimental research programs *per se*. It is, however, a clearinghouse for information and technology related to cleanup of nuclear waste and environmental remediation.

Member States participate in sharing information on cleanup technologies, remediation planning, and site characterization through IAEA as well as in using technical and safety information and recommendations contained in the IAEA publications. Other forms of information dissemination and capacity building include workshops, training courses, scientific visits, and expert meetings (supported by the Department of Technical Cooperation).

The expectation is that Member States will eventually have in place a proper infrastructure and technologies for managing their radioactive legacies and resolve all related issues in a timely, safe, and cost-effective manner. Partnering with the IAEA includes both being a contributor of technology as well as a user of best practices developed in other countries. Toward this objective, the IAEA is the creator of the Network of Centers of Excellence in Environmental Remediation (ENVIRONET) that, in conjunction with other networks recently created by the IAEA (e.g., International Decommissioning Network [IDN]; and Waste Disposal Network [DISPONET]), will expedite the exchange of information and help to disseminate good practices in this field.

Nuclear Energy Agency of the Organisation for Economic Co-operation and Development

The mission of the Nuclear Energy Agency (NEA) is to assist its member countries in maintaining and further developing, through international cooperation, the scientific, technological, and legal bases required for the safe, environmentally friendly, and economical use of nuclear energy for peaceful purposes. The NEA works as:

- A forum for sharing information and experience and promoting international cooperation, and
- A center of excellence that helps member countries to pool and maintain their technical expertise.

The NEA's 30 member countries house about 85 percent of the world's nuclear power capacity.

Hans Riotte, head of Radiation Protection and Waste Management within the NEA, stated that the NEA has long been, and continues to be, a leading organization in the field of radioactive waste management and, in particular, geologic disposal (Riotte and Nokhamzon 2008). The NEA's Radioactive Waste Management Committee addresses all aspects of radioactive waste management, including developing public confidence. The Committee on Radiation Protection and Public Health addresses future directions for radiation protection policy and operational radiation protection, including planning to deal with nuclear emergencies and reducing occupational exposures.

Jean-Guy Nokhamzon, chair of the NEA Cooperative Programme on Decommissioning, described 20 years of exchange of information about D&D projects among the program's participants. The projects have included 29 research and nuclear power reactors and 13 fuel cycle facilities. Twenty-four organizations from 12 countries have participated. Information exchanged has included:

- Use of remote systems and robotics,
- Partial dismantling of plants,
- Dismantling large components, and
- Release of alpha-contaminated areas.

Nokhamzon stated that member countries gain benefit of earlier experience and spread it on a larger scale. They share in the development of improved technologies for dismantling and demolition and encourage continued research on new technologies. Current technologies have demonstrated their effectiveness and robust performance in numerous decommissioning activities. The dissemination of best practices and sharing of information in international workshops and conferences has proven to be a good basis for an effective cooperation. He observed that international cooperation is important for meeting future cleanup challenges.

Federal Partnership Programs: The Role of the Small Business Innovation Research Program

In various forms, the federal government has been partnering with both the private sector and academia since its early history. The late Professor Vernon Ruttan, one of the world's leading development economists, with more than 50 years of distinguished academic and nonacademic experience in the United States, is quoted in his recent book: "Government has played an important role in the technology development and transfer in almost every U.S. industry that has become competitive on a global scale" (Ruttan 2001 p. 602). The government has used a variety of partnering models from direct support of basic research through the National Science Foundation, to multiparty arrangements involving the government, private sector, and academia as co-partners in developing technology.

Charles Wessner, a National Academies scholar and the director of the National Research Council's (NRC) program on Technology, Innovation, and Entrepreneurship overviewed the role of the Small Business Innovation Research Program (SBIR) in facilitating partnerships between federal laboratories, universities, and the private sector. The goals of these partnerships have been to:

- Stimulate technological innovation,
- Use small businesses to meet federal R&D needs,
- Increase employment,
- Foster and encourage participation in technological innovation by minorities and women, and
- Increase private-sector commercialization of innovations derived from federal R&D.

As the SBIR program approached its 20th year of operation, the U.S. Congress asked the NRC to conduct a comprehensive study of how the program has stimulated technological innovation and used small businesses to meet federal R&D needs and make recommendations on improvements to the program.[12] In the past 5 years, the NRC has published a series of reports based on comprehensive studies that were carried out on the SBIR program. Over 7,000 projects were surveyed and about 100 case studies were conducted. Regarding the program itself, the studies concluded that:

- The program is sound in concept and effective in operation;
- Twenty percent of the participating companies were created because of the program;
- Nearly 60 percent of the participants reach the market in some fashion; and
- The program creates greater choice for federal procurement, adding new options and increasing competition.

The studies also concluded that small U.S businesses:

- Are key players in bringing new technologies to market;
- Generated 60 to 80 percent of net new jobs annually over the last decade;
- Employed 39 percent of high-tech workers, such as scientists, engineers, and computer workers;
- Produced 13 to 14 times more patents per employee than large firms and the patents are of high quality, being twice as likely to be among the top 1 percent of patents cited; and
- Are a key source of innovation for themselves and for large companies.

Currently a $2.3 billion-per-year program, the SBIR is the largest U.S. partnership program. Its participants include all federal agencies with $100 million or larger R&D budgets, such as the Departments of Agriculture, Commerce, Defense, Energy, Labor, Housing and Urban Development, and Veterans Affairs. The success of the SBIR has led a number of foreign countries, including Finland, India, Japan, Korea, Netherlands, Russia, Sweden, Taiwan, and the United Kingdom, to adopt various aspects of the program in the support of development within their own countries. Clearly the SBIR program is well suited to provide opportunities for EM to create private–academic partnerships to efficiently leverage its R&D.

---

[12] See SBIR Reauthorization Act of 2000 (H.R. 5667, Section 108.

COMMITTEE OBSERVATIONS ON THE ROLE OF ROADMAPPING AND LEVERAGING IN R&D

EM has been pursuing new technologies to improve its site cleanup efforts and to leverage its technology development and implementation with other organizations for a number of years. However, according to presentations and other information received by the committee, much of this effort is perceived as having achieved only limited success in providing new technologies to make cleanup "faster, cheaper, and safer," which was the mission of EM's original Office of Technology Development, established in 1989. The most notable success, which was described at all of the committee's site visits, was the development of the caustic-side solvent extraction process for the SRS. This process is being implemented and is expected to provide a major improvement in the removal of cesium in SRS salt waste processing (NRC 2000a, 2001a).

In EM site cleanup programs, contractors fund R&D to address their immediate cleanup needs. However, science and technology development that is necessary to provide transformational technology (i.e., technology that could provide a breakthrough in risk reduction, cost, or schedule) typically requires a longer time horizon than a contract or contractor can include within a particular task. In its interim report the committee observed that the responsibility for providing sustained R&D support that can lead to transformational cleanup technologies resides at the EM headquarters level (Appendix H).

Despite the increasing scope and complexity of EM's responsibilities for cleanup, investment in medium- to long-term R&D in support of these activities has been decreasing (Chapter 1). The perception that EM's technology development efforts have had limited success in providing new technologies for the cleanup program suggests that the efforts can be made more effective. In particular, the efforts can be improved by:

1. Improving the Roadmap and using it as a central tool for EM's R&D planning and for communicating its plans and programs to other organizations, including other DOE offices, federal agencies, and Congress; and
2. Better application of the basic principles of leveraging research, with recognition that legacy waste cleanup is a national responsibility that requires other organizations to willingly partner with EM.

Improving the EM Roadmap

The technology roadmapping process has been used widely as a planning tool in industry and government to match technology resources with desired product or process outputs. EM has begun to develop such a roadmap to reduce technical risk and uncertainty in its cleanup program. The term "roadmap" is not accidental. Like a true roadmap, a technology roadmap gives details of how to get to the destination. Furthermore, if parts of the trip (getting there) require using transportation provided by different carriers (partnerships) the roadmap will spell out such details based on agreements with all of the partners.

"Research that creates the right technologies at the right time is critical to competitive success in many industries." The committee judges that this statement by Christensen et al. (2004, p. 1) also applies to the tasks that face EM in the cleanup program. Investments in R&D eventually have to pay off or they are wasted opportunities. EM's payoffs from R&D will vary from better knowledge and understanding of the cleanup issues to actual cleanup capabilities that were not available prior to the work supported by the R&D investment.

A roadmap lays out medium- and long-term R&D plans that will allow EM to effectively communicate expected payoffs from the research investment to its investors (Congress, DOE management, and the public) and its implementers (DOE management, partnering agencies, and contractors). All of the essential steps of the roadmapping process and in particular the schedule of R&D investments and expectations, such as those steps discussed with the committee by Scouten (2008) and described in his publication with Cosner et al. (2007), are important elements of a roadmap that can be effectively used as a planning and implementation tool. A roadmap can be an especially effective tool in planning and communicating expectations from R&D activities that are leveraged in partnerships.

Applying the Principles of Leveraging

To be successful in leveraging R&D, all participants in the collaboration must receive benefits from the partnership in order for it to be sustained. Moreover, all participants in the collaboration should bring something to the partnership that is needed by the other partners in the collaboration. This may include financial resources or specific expertise that other partners can build on and benefit from.

In the plenary session of the committee's April 2008 meeting on R&D leveraging, Mike Dalton summarized the best practices for leveraging R&D from years of experience in working with private-sector companies in developing successful partnerships (Dalton 2008). The process to develop effective leveraging partnerships is not complicated; however, it requires discipline to be carried out in rigorous detail that does not leave any of the essential elements undefined.

The four steps in the particular process that was used in the Los Alamos example—see Sidebar 4.2 and Slowinski and Sagal (2003)—are:

1. **Want:** What external resources do we need to succeed in our mission?
2. **Find:** What mechanisms will we use to find these resources?
3. **Get:** What processes will we use to plan, structure, and negotiate an agreement to access the resources?
4. **Manage:** What tools, metrics, and management techniques will we use to implement the relationship?

There is nothing revolutionary in this process; however, by spelling out the details of each step and following through in implementing them, the process takes on a rigor that minimizes the risk that some critical step in developing and implementing the partnership results in its failure.

As applied to the EM task, Step 1 in the above process is clearly the EM roadmap that is under development and is being shared with potential partners. At this step in the process, the Roadmap can be developed in conjunction with potential partners and include time lines for R&D to address technology gaps. Step 2, identifying partners to share resources, was described in the first part of this chapter with some examples of opportunities for partnerships. More comprehensive mechanisms involving the technical community of which EM is a part can also be used to search out the collaborative opportunities. Step 3 should be a formalized, structured agreement in which partners in the leveraging activity have an understanding of what they bring that others in the partnership need and what they expect to take away, so that all partners buy into a negotiated agreement. Finally, Step 4 is the monitoring of progress toward the success goal using metrics that have been agreed upon by all partners and are appropriate to the task at hand.

As simple as these techniques sound, they are not easy to implement given the focus, orientation, and organizational objectives of the partners. This is precisely why a rigorously adhered-to process is required to ensure that, despite the individual cultural and political drivers that each organization brings to the partnership, the specific goals of the leveraging partnership are achieved.

SUMMARY AND CONCLUSIONS ON LEVERAGING

Successful leveraging of the R&D that EM needs to address its technology gaps requires: (1) a roadmap that not only spells out the details of technology gaps that are obstacles to the EM cleanup but also a time line that allows both investors and implementers to understand how the plan meets the technology needs and milestones for site cleanup; and (2) an execution process for partnering that is rigorous and transparent to all leveraging partners.

The organizations that attended the committee's April meeting are, in the committee's judgment, many of the ones that can provide good opportunities for mutual leveraging of R&D with EM. Many are already doing so. Clearly other opportunities for partnering, especially with organizations outside of the federal sector, will open as EM further develops and implements its Roadmap.

In the planning and development of its Roadmap, details, time lines, and close interactions with potential leveraging partners can help ensure that there are viable connections between EM's roadmapped objectives and the support it can negotiate with these partners. The committee wishes to reemphasize the necessary *quid pro quo* nature of these partnerships and the need to ensure that EM is fully vested to enter into such relationships as an equal partner.

5

Findings and Recommendations

To help sustain future investments for research and development (R&D) of new technologies for Department of Energy (DOE) site cleanup, Congress requested that the DOE Office of Environmental Management (EM) develop an Engineering and Technology Roadmap.[1] As EM began work on the Roadmap, the DOE Assistant Secretary for Environmental Management and the EM Office of Engineering and Technology (OET) turned to the National Research Council (NRC) for assistance. The NRC in turn empaneled the committee that prepared this final report. The preceding chapters have addressed the four items of the committee's statement of task.[2] This chapter summarizes the committee's advice to EM and presents its formal findings and recommendations.

This chapter's first section recalls the committee's key observations in its interim report (reprinted in Appendix H) and, from these observations, develops overarching themes relevant to the EM cleanup mission and to the EM roadmap. The second section presents the committee's findings and recommendations. The third section gives a final set of observations that may lead to an enhanced role for engineering and technology development in assisting EM to successfully complete its 30-year DOE site cleanup mission.

---

[1] EM's Engineering and Technology Roadmap (DOE 2008b) is referred to as the EM roadmap or simply as the Roadmap.

[2] The committee's Statement of Task appears in Chapter 1, Sidebar 1.1.

OBSERVATIONS FROM THE INTERIM REPORT AND OVERARCHING THEMES

Three key observations from the interim report helped frame the committee's deliberations that led to this final report. These observations and the overarching themes that emerged from them may help guide EM's future R&D roadmapping and enhance its cleanup efforts.

Observation 1: The complexity and enormity of EM's cleanup task require the results from a significant, ongoing R&D program so that EM can complete its cleanup mission safely, cost-effectively, and expeditiously.

The EM cleanup program will continue for at least another 30 years and may cost \$300 billion. This is clearly a long-term commitment for the U.S. government and for DOE. Considering how science, technology, and the focus of public debate on science and technology issues have changed in the past 30 years, EM will face continuing changes in the technical state of the art and in public expectations as it pursues its cleanup program.

Three overarching themes emerged from this study that bear on the long-term nature of the EM program: (1) establishing cleanup goals (i.e., deciding how clean is clean); (2) gaining and preserving knowledge for cleanup; and (3) striking an appropriate balance between long-, medium-, and short-term R&D. Incorporating these overarching themes in the Roadmap can help ensure that EM has sound scientific and technical bases for its long-term planning and decision making.

Establishing Cleanup Goals

The need to decide "how clean is clean" is pervasive in the EM cleanup program. The results of such decisions are primary drivers for technology initiatives in the main program areas in the Roadmap, including retrieval of tank waste residues, groundwater and soil remediation, and end points for facility deactivation and decommissioning. Deciding how clean is clean is continually evolving among DOE, its regulators, and public citizens.

One example of evolving cleanup objectives is determining the degree to which tank waste heels must be removed in order that the residue can be defined as "waste incidental to reprocessing," which can be left on DOE sites, rather than defined as "high-level waste," which must be disposed in a specially licensed geologic repository. Technical factors that may affect these decisions were examined in NRC (2006b). The incidental waste provisions in law apply only to Idaho and Savannah River wastes; how the law or similar provisions might be applied to Hanford tank waste is yet to be determined. At every site, future decisions about when to declare a waste tank to be clean enough for permanent closure will be made in negotiations among DOE, its federal and state regulators, and public representatives.

Another example of evolving cleanup objectives is the agreement finalized on July 1, 2008, by the state of Idaho and DOE, which requires DOE to remove most of the transuranic (TRU) waste buried in the Subsurface Disposal Area at the Idaho site. Initially DOE considered the waste, buried between 1952 and 1970, to be permanently disposed. DOE recognized in 2001 that some buried TRU should be removed (DOE 2001; NRC 2002). Six years of litigation, technical analyses, and negotiations begun in 2002, when the state brought legal proceedings in U.S. District Court to determine DOE's responsibility for these wastes, eventually culminated in the 2008 agreement.[3]

To support decisions on how clean is clean, DOE typically makes a technical assessment of a cleanup problem that balances the risks, environmental impacts, costs, and schedules for varying degrees of cleanup. For the assessment to be acceptable and defensible requires that the assessment and the process to prepare it be transparent to stakeholders (including the public), be scientifically sound, and consistent with assessments of similar problems. The examples above indicate that preparing acceptable and defensible assessments is not easy for DOE, and that DOE may have difficulty in defending its decisions. A necessary role for ongoing EM R&D is to provide state-of-the-art knowledge to guide and support its cleanup decision making as society's perspectives on how clean is clean continue to evolve.

Gaining and Preserving Site Knowledge for Cleanup

In addition to EM R&D's providing state-of-the-art knowledge for decision making, success of a 30-year program requires the preservation of knowledge and expertise. This involves, first, maintaining a core of personnel who have first-hand knowledge of site cleanup needs, how they have developed, and previous lessons learned; and, second, managing accumulated information and knowledge.

Knowledgeable Personnel

Few who are active today will see the end of the site cleanup program, and more to the point, many of today's site-knowledgeable personnel are retiring. Over the long haul of the cleanup program, personnel who can develop expertise, experience, and understanding of a site's history must be recruited and retained. Continuity in the specialized expertise and experi-

[3] See http://www.deq.idaho.gov/inl_oversight/contamination/agreement_waste_removal_2008.cfm.

ence base required for the cleanup effort is an essential ingredient for the success of EM's long-term program. Knowledgeable personnel can provide historical insights concerning which cleanup approaches were effective—or not—and how to efficiently and safely conduct operations at facilities that have evolved over half a century. The result of not having such knowledge is likely to be higher cost resulting from ignorance of lessons learned, leading to additional resources being needed to relearn what had been known and compensate for mistakes made through ignorance. Inefficient or potentially unsafe operations are also likely results of such ignorance.

All four sites visited by the committee cited challenges in maintaining continuity in technical staff with the experience required to analyze, plan, and implement activities to clean up the DOE sites. Challenges are being encountered in recruiting technical personnel ranging from technicians performing cleanup operations and supporting R&D to Ph.D.-level staff performing science, technology, and design activities. Experienced personnel at all of these levels are retiring and replacements are either not being made or are difficult to attract to this work. These challenges were attributed to a combination of factors: unstable budgets, increasing demand for nuclear-savvy personnel in the resurgent civilian nuclear sector, and a perception that career opportunities would be short-lived because the cleanup program was going out of business in the near future. In cutting back EM R&D funding, DOE has, perhaps inadvertently, sent a message that says "cleanup does not need new technology" and "cleanup is not as important as other programs within DOE." Messages such as these have an impact on recruiting and retaining the necessary personnel, as well as on how the EM organization is viewed from both inside and outside of DOE.

Knowledge Management

Chapter 3 stated that maintaining information archives to house and integrate the growing knowledge of site characteristics that govern contaminant fate and transport is an important role for the national laboratories. More broadly, EM and the national laboratories have the opportunity, and the responsibility, to maintain the accumulated knowledge from site cleanup programs and R&D in ways that are easily accessible. The term "knowledge management" refers to a discipline that seeks to improve the performance of individuals and organizations by maintaining and leveraging the present and future value of knowledge assets. It includes human and automated activities as well as the processes an organization uses to optimize its intellectual capital to achieve organizational objectives.[4]

One element of government reform efforts initiated by the President in

[4] See http://www.systems-thinking.org/kmgmt/kmgmt.htm.

2001 is expanding electronic government and one component of the E-government effort is knowledge management.[5] As a consequence, knowledge management has been getting increasing attention during the last several years under the leadership of the Best Practices Committee of the Federal Chief Information Officer Council (FCIOC) of which DOE is a member.[6] The resources being organized by the FCIOC and collaboration with other FCIOC members could be valuable to the EM cleanup program. Regular symposia on knowledge management are organized by the Digital Government Institute. The National Aeronautics and Space Administration, an agency supporting substantial R&D efforts and having complex operational issues conceptually similar to the EM cleanup program, has an active knowledge management program.[7] The Roadmap can be an important component of knowledge management by using the R&D needs it displays as a basis for identifying high-priority areas of knowledge on which knowledge management efforts could focus.

Balance Between Long-, Medium-, and Short-Term R&D

It was clear from the committee's visits to all four sites that the research being done was largely driven through the cleanup contractors who primarily supported short-term research to meet the contractors' objectives. However, the long period of time over which EM will be performing increasingly difficult cleanup tasks and the nature of some of the technology gaps indicate that a portfolio of longer-term R&D programs is needed. The scope of the medium- to longer-term research would include addressing the gaps that this committee identified, knowledge building, and seeking transformational technologies. One would expect the Roadmap to include a mix of R&D projects that span a spectrum of time in terms of availability of results. As stated in the interim report (Appendix H), providing support for the longer term is a responsibility of DOE.

Observation 2: By identifying the highest cost and/or risk aspects of the site cleanup program, the EM roadmap can be an important tool for guiding DOE headquarters investments in longer-term R&D to support efficient and safe cleanup.

As it continued its deliberations after issuing the interim report, and especially after considering the information presented at its April 2008 meeting on leveraging EM investment, the committee concluded that road-

[5] See http://www.whitehouse.gov/omb/inforeg/egovstrategy.pdf.
[6] See http://www.cio.gov/index.cfm?function=aboutthecouncil.
[7] See http://wiki.nasa.gov/cm/wiki/?id=1926).

mapping engineering and technology for the remainder of EM's program is essential for its success. While EM's draft Engineering and Technology Roadmap, issued in April 2007, and the substantially similar finalized version, issued in March 2008, describe in some detail the technical risks and uncertainties and the types of strategic initiatives that are needed to address them,[8] the committee judged that there are significant opportunities to improve the Roadmap.

Notably the EM roadmap provides no time lines for its initiatives or connections between the initiatives and EM site cleanup milestones. This is rather like drawing a map by simply listing cities without placing them geographically on the map or showing highway interconnections. A much more useful EM roadmap will show when and how the initiatives address technology gaps such as those identified in this study. This could be compared to deciding the detail of how a trip will be taken, by which roads, and on what schedule. The importance of this is to be able to plan the internal as well as external programs needed to address these gaps. Without this planning the result of a great deal of R&D can be a random walk that will not lead to timely new technologies.

Once such planning is done, it needs to be communicated to those responsible for supporting the program (DOE management and Congress) and then to the potential community of partners who will participate in helping to execute the necessary R&D and cleanup operations. Implementation of the Roadmap will require effective management and monitoring of the R&D programs to meet the milestones laid out in the Roadmap, or if those turn out to be unachievable, to alter the Roadmap accordingly.

EM's OET does not have the resources necessary to sustain all of the capabilities described in Chapter 3 that are necessary for its R&D work. As a consequence, if EM's R&D needs are to be met, a large portion of the R&D work that is needed by EM will involve partnering with other organizations. As discussed in Chapter 4, successful partnering will follow a detailed and rigorous process from planning to implementation in order to be successful.

Observation 3: The national laboratories at each site have special capabilities and infrastructure in science and technology that are needed to address EM's longer-term site cleanup needs. The EM roadmap can help establish a more direct coupling of the national laboratories' capabilities and infrastructures with EM's needs.

[8] For this study the committee was charged to identify "science and technology gaps." The EM roadmap uses the terminology "technical risk and uncertainty" to describe similar obstacles to EM's cleanup work.

Supporting the DOE sites' nuclear material production involved cutting-edge science from the beginning of the nuclear age until well into the Cold War era. The national laboratories have continued to produce world-class science, but in doing so they have moved away from their roots in DOE site support. The special capabilities of the national laboratories necessary to sustain longer-term EM R&D, described in Chapter 3, are now mostly supported by non-EM programs (Appendixes D, E, F, and G). The movement away from EM work was clear to the committee during its site visits. As one example, Roberto (2007) stated that Oak Ridge National Laboratory now applies its strengths in science and technology to six major missions: neutron sciences, ultrascale computing, advanced materials, systems biology, advanced energy systems, and national and homeland security. The relation between these missions and EM cleanup needs seems tenuous at best.

In its earlier discussion about EM's need to retain the capabilities of knowledgeable personnel, the committee noted that beginning in about 2001, EM, perhaps inadvertently, sent a message that cleanup does not need new technologies. Approximately coincident with EM's withdrawal of R&D support, the national laboratories developed new sponsors for cutting-edge science. DOE moved the EM Science Program to the Office of Science (SC) in 2002.

After visiting the national laboratories, hearing from SC, and reviewing their literature, the committee concluded that EM-related research is out of the laboratories' mainstream. Except for a relatively few researchers who are still engaged in EM problems (e.g., tank waste chemistry, new process development) most research personnel consider other scientific areas to be more exciting and higher profile. There is obviously a synergistic relationship between prospects for sustained research funding in a given area and the engagement of researchers in that area.

In spite of this current rather negative perception of EM's place at the R&D table, there are bright spots. The SC's Environmental Remediation Science Program remains strongly focused on DOE-relevant contaminants and their fate and transport in the subsurface. In July 2008, 12 scientists representing the four national laboratories visited by the committee and OET prepared a detailed report on *Scientific Opportunities to Reduce Risk in Nuclear Process Science*. The summary of this report states, "Over the last 3 years, DOE's Office of Environmental Management (EM) has experienced a fundamental shift in philosophy. The mission focus of driving to [site] closure has been replaced by one of enabling the long-term needs of DOE and the nation" (Bredt et al. 2008, p. iii).

Renewed national interest in nuclear energy and advanced fuel cycles will provide increasing opportunities for synergy among EM, other DOE offices, and the private sector. Investments by EM for R&D and maintaining site capabilities can be leveraged to also support the management of newly generated wastes, while investments in nuclear capabilities and infrastructure by EM's partners will help support EM's technology development. By clearly roadmapping its science, engineering, and technology programs EM has the opportunity to become better integrated with the rest of DOE; to reengage the capabilities of the national laboratories, universities, and the private sector; and to maintain access to these capabilities for the duration of the cleanup program.

FINDINGS AND RECOMMENDATIONS

The Statement of Task directed the committee to provide findings and recommendations, as appropriate, to EM on maintenance of core capabilities and infrastructure at national laboratories and EM sites to address its long-term, high-risk cleanup challenges. In carrying out its task, the committee judged that EM's Engineering and Technology Roadmap can be a key tool to ensure that core capabilities and infrastructure remain available to EM over the next 30 years of the site cleanup program.

The committee provided findings and recommendations in two areas: (1) improving the Roadmap so that it clearly demonstrates the role of R&D in the EM cleanup mission; and (2) establishing R&D programs that utilize national laboratory, site, and private-sector capabilities to bridge the science and technology gaps identified in Chapter 2.

Improving the Roadmap

FINDING: The Roadmap is an important and much needed tool for guiding DOE headquarters investments in longer-term R&D to support efficient and safe cleanup.

FINDING: The current Roadmap describes technical risks in the EM site cleanup program and R&D initiatives to mitigate these risks. However, it does not connect these initiatives to major milestones in the EM cleanup program.

RECOMMENDATION 1: **EM's Office of Engineering and Technology should update its 2008 Roadmap to include performance metrics and time lines for accomplishing its R&D initiatives to ensure that results are useful and timely to meet EM's site cleanup milestones.**

RECOMMENDATION 2: **The DOE Assistant Secretary for Environmental Management should require periodic, future updates of the Roadmap to ensure that it remains current with major mid- to long-term milestones in**

the cleanup program. At a minimum, the Roadmap should be updated at least every 4 years at an appropriate time to help ensure carryover of programs and their rationales into new administrations.

The preceding findings and recommendations are elaborated in Observation 3 earlier in this chapter.

FINDING: EM is the DOE office designated to clean up the nuclear materials production sites of the Cold War. Cleaning up these legacy sites nevertheless remains a responsibility for all of DOE and the nation. EM cannot complete its mission without the active cooperation of other DOE offices and federal agencies. The Roadmap can be improved by specifying opportunities for cooperative work with the national laboratories and other DOE and federal agencies.

Examples of such leveraging opportunities were detailed in Chapter 4.

RECOMMENDATION 3: **The EM Office of Engineering and Technology, with support of the Secretary of Energy and the Assistant Secretary for Environmental Management, should engage other federal organizations (e.g., Department of Defense, Department of Homeland Security, Environmental Protection Agency) and DOE offices (e.g., Office of Science, Office of Nuclear Energy, Office of Legacy Management) to specify Roadmap intersections with the others' R&D programs to ensure that opportunities for joint work are recognized and implemented in timely fashion to produce results that are useful to EM.**

This could be done by convening workshops at which participants exchange information on their cleanup-relevant R&D programs and milestones. The OET did this to a limited extent in preparing the 2008 Roadmap. The workshops could be arranged to provide timely information for periodic updates of the Roadmap according to Recommendation 2. Other mechanisms to identify leveraging opportunities include attending program reviews of other organizations, literature reviews to identify organizations and individuals working on topics of interest to EM, and requests for expressions of interest on federal procurement websites.

RECOMMENDATION 4: **The DOE Assistant Secretary for Environmental Management and the Office of Engineering and Technology should use the Roadmap as a primary means of communicating EM's technology needs, R&D planning, and accomplishments within DOE to other federal and state agencies, and ultimately to Congress.**

FINDING: The scientific and technical state of the art will evolve during the next 30 years of the EM site cleanup program, as will public expectations for the cleanup goals. A robust EM science, engineering, and technology program will be required to keep up with these evolutions, to provide up-to-date bases for EM's cleanup decisions, and to maintain a skilled workforce. Such a program would consist of short-, medium- and long-term components that address near-term as well as longer-term goals in the program. Presently it appears that only short-term goals are being addressed through contractor-supported R&D and issues that require a longer time line are not being addressed.

RECOMMENDATION 5: **EM and its Office of Engineering and Technology should include in its Roadmap the overarching themes of (1) maintaining state-of-the-art cleanup objectives as science, technology, and the public's expectations evolve during the next 30 years; (2) maintaining and distributing up-to-date knowledge resources relevant to site cleanup; and (3) developing a balanced R&D portfolio that addresses short-, medium-, and long-term issues.**

In the first instance, the Roadmap might identify the organizations responsible for providing technical data and timely R&D milestones to support key EM site cleanup decisions (e.g., the cleanup objective for a waste burial ground, a groundwater plume, or a decommissioned facility). In the second instance, the Roadmap might include objectives for hiring and retaining personnel, and for information archiving, at specified milestone times during the next 30 years.

Bridging EM's Science and Technology Gaps

FINDING: The unique chemical, physical, and radiological properties of waste and contamination at the EM cleanup sites and the unique subsurface characteristics of the sites themselves require special capabilities of the sites and their associated national laboratories to sustain long-term R&D for EM's 30-year cleanup program. These special capabilities include qualified, experienced personnel and facilities for radiochemical, engineering, and field experiments. It is Congress's and DOE's responsibility to maintain the national laboratories' capabilities not only for cutting-edge scientific research, but also for research applied to national problems such as DOE's Cold War legacy cleanup.

In Chapter 2 of this report the committee identified science and technology gaps that may impede EM's cleanup program. In Chapter 3 the committee identified special capabilities of the sites and national laboratories that will be needed for R&D to bridge the gaps. In Chapter 4 it discussed ways that EM could better leverage its R&D. In roundtable discussions during the final synthesis of its work, the committee noted that a number of scientific disciplines or areas of investigation are important to multiple EM high-priority R&D needs. These were crystallized into nine targeted R&D programs that are described in the next recommendation.

***RECOMMENDATION 6:* The EM Office of Engineering and Technology, with support from the Secretary of Energy and the Assistant Secretary for Environmental Management, should lay out in its Roadmap programs that include research in the following:**

- **Radiochemistry of EM wastes and contaminants;**
- **Long-term performance of cementitious materials;**
- **Retrieval technology for high-level waste;**
- **Alternative and advanced waste forms and production methods;**
- **Rheology of waste sludges and slurries;**
- **Long-term behavior of in-ground contaminants;**
- **Advanced sensors, detectors, and data transmission technology for subsurface monitoring;**
- **Advanced near-surface engineered barrier systems to control contaminant release to the environment; and**
- **Surface characterization of solid materials.**

These research programs are discussed below.

Radiochemistry of EM Wastes and Contaminants

As noted earlier, the essential uniqueness of the EM cleanup program is processes and legacy facilities and materials containing substantial amounts of a wide variety of radionuclides. Essentially all of the technology gaps described in Chapter 2 involve radiochemistry either as related to radioactive waste or to radioactive contamination in the environment. Some examples of this are as follow:

- Waste processing: Processing of tank waste at the Savannah River Site (SRS) and Hanford involves a number of processes (solvent extraction, precipitation, and ion exchange) for separating various radionuclides in which the chemistry of the radionuclides is central. Additionally, the effects of radiation on materials during waste processing (e.g., degradation of organic reagents, production of hydrogen by radiolysis) are also important.
- Subsurface contamination: Radionuclides have leaked or been released into the subsurface at all four sites visited by the committee. The

identification and design of remedial technologies requires a firm understanding of the waste and the biogeochemistry of the target contaminants.

- Deactivation and decommissioning: These activities often involve the need to remove radionuclides from surfaces of nonporous materials or the matrix of porous materials. Chemical or electrochemical techniques are often useful in this application, and creating such techniques requires knowledge of the chemistry of the target radionuclides.

There are radiochemistry capabilities at the national labs, academia, and medical institutions. However, (a) the expertise is mostly focused on a relatively narrow set of radionuclides many of which are not relevant to EM, and (b) there is relatively little emphasis on situations relevant to EM such as radionuclide chemistry in alkaline solutions, in groundwater and soils, or on surfaces. EM needs to maintain and expand its base of knowledge and expertise in radiochemistry related to the isotopes at its cleanup sites as a resource for the ongoing cleanup effort.

Long-Term Performance of Cementitious Materials

Cementitious materials have been or will be used in many applications relevant to EM cleanup program activities concerning high-level waste (HLW) and nuclear material management, groundwater and soil remediation, and facility deactivation and decommissioning. Examples include:

- Use of engineered[9] grouts to fill tanks and pipelines from which waste has been retrieved at the Hanford, Idaho, and Savannah River sites;
- Use of engineered grouts to stabilize low-activity wastes (LAW) resulting from tank waste processing;
- Injection of grouts into the subsurface to stabilize legacy waste disposals and provide a barrier to contaminant migration;
- Use of concrete for surface structures (e.g., pads, vaults) for disposal of LAW, and for waste emplacements (e.g., concrete caissons); and
- Injection of grouts into the interstices of rubbleized concrete structures (e.g., production reactors, reprocessing plants) to stabilize residual hazardous materials.

The long-term performance of cementitious materials is important in all of the above applications. Chapter 2 describes gaps concerning the need for improved understanding of the behavior of cementitious materials that have been mixed with wastes or that are used as barriers to confine radionu-

[9] Engineered grout is designed to maintain chemical conditions such as high pH and low E_h that impede radionuclide dissolution and transport.

clides. These gaps include WP-1 and WP-2 (grouts to fill nearly empty HLW tanks and stabilize LAW), GS-3 (grouts injected to stabilize waste burial sites or constitute subsurface barrier walls), and DD-3 (decontamination of concrete surfaces).[10]

The applications in which cementitious materials are used involve a range of natural, engineered, and mixed systems as well as a variety of cementitious materials. The cementitious materials often have common features, such as portland cement as the principal ingredient and degradation driven by water, oxygen, carbon dioxide, and certain species normally present in low concentrations.

Retrieval Technology for HLW

Gap WP-1 described the need for waste retrieval technologies that leave less residual waste in a tank and cost less to implement. This need may be especially acute for tanks containing debris, Hanford tanks that have leaked, and SRS tanks that contain cooling coils. EM relies on large pumps to sluice and remove the bulk of the waste from a tank followed by other devices including minibulldozers, end effectors such as scabblers and high-pressure water jets deployed on crawlers or articulated arms, and devices similar to a carpet steam cleaner to remove most of the remaining waste. All of these devices have antecedents in applications unrelated to EM (e.g., homeland security, improved manufacturing efficiency, advanced battlefield approaches) and substantial relevant R&D continues to be supported by other organizations.

However, EM's retrieval activities have a number of unique features that are unlikely to be taken into account by the R&D of other organizations: high levels of radiation, small access openings, high temperatures, harsh chemicals, and internal tank structures that can obstruct retrieval devices. As a consequence, EM needs a toolbox of technologies that can be applied to any given retrieval challenge, and it needs to continually develop and adapt new technologies as the state of the art advances.

Alternative and Advanced Waste Forms and Production Methods

The production of borosilicate glass waste forms is the most expensive activity in the EM cleanup program and it drives the duration of the program. Gaps WP-4 and WP-5 in Chapter 2 describe the need to develop advanced waste forms that have higher waste loadings, advanced production methods that have a higher throughput, treatment methods that remove interfering components that either compromise performance or

[10] The gap numbers refer to the gaps set forth in Chapter 2.

impair production, and alternatives that could avoid the need for high-temperature processes. Other organizations conducting fuel reprocessing have advanced technology for production of borosilicate glass. Alternative approaches (e.g., steam reforming) have been developed for purposes other than radioactive waste management. However, EM's applications have unique features related to the chemistry of the waste, the presence of large amounts of nonradioactive chemicals, the heterogeneity of the waste due to the multiplicity of processes that produced it, and the need to adapt alternatives to work in a radioactive environment.

Rheology of Waste Sludges and Slurries

Both Hanford and SRS have millions of gallons of viscous sludge to be mobilized, transported through pipes, stored in holding tanks, and eventually remobilized for processing. Additionally, as discussed in Chapter 3, pilot-scale test facilities require the use of simulants that accurately mimic the physical or chemical properties of the waste. Gaps WP-1 and WP-3 relate to this issue.

Failure to accurately predict the flow properties of slurries is a leading cause of process failure. The need to transport and process sludges, and to prepare sludge simulants is not unique to EM nor is the capability to do R&D on these topics. Industries ranging from plastics to concrete routinely transport and process non-Newtonian solids and develop appropriate simulants. Additionally, in the last decade or so, industry and national laboratory researchers have made substantial advances in computational approaches to predicting fluid properties. However, this work does not focus on the specific materials that EM must process, the range of compositions these materials have, and complications such as hydrogen generation resulting from radiolysis. This leads to the need to extend and adapt new knowledge of rheology, chemistry, and radiolysis to the specific materials and processes relevant to EM.

Long-Term Behavior of Inground Contaminants

Currently available technologies, including EM's baseline technologies, are insufficient to remediate many of DOE's groundwater and vadose zone contaminants (gaps GS-1 and GS-2). In addition, DOE's cleanup plan includes leaving some contamination in the subsurface (gap GS-3). These gaps identify a need to locate, understand, and predict the long-term behavior of inground contaminants as an essential component of the successful application of both barrier and "cocooning" strategies. However, current technologies and approaches to characterizing, conceptually understanding, and modeling subsurface properties and processes are both inefficient and insufficient, and can lead to unreliable predictions of subsurface contaminant behavior.

In addition to change in the (geo)chemical form of a contaminant over time, temporal changes in the groundwater system can also affect contaminant mobility. For example, elimination of process water infiltration at the Hanford site has led to a declining water table, thus "thickening" of the vadose zone. Contaminants remaining in this zone now reside in an altered hydrologic and geochemical context. Another example is provided by changes in groundwater chemistry that will also evolve with time. An aqueous solution that was acidic when released may become neutralized through hydrologic dilution and flushing processes or reactions with the aquifer matrix in the subsurface. Saline solutions may become concentrated through evaporation in the shallow vadose zone or diluted through mixing in the saturated zone. These examples illustrate the contextual system changes that can affect the temporal behavior of inground contaminants.

An ability to reduce uncertainties in predictions of long-term behavior of inground contaminants through improvement in technologies and approaches to characterizing, conceptually modeling, and predicting the critical hydrologic, geochemical, and biogeochemical processes affecting DOE's inground contaminants at appropriate spatial scales is important for EM's cleanup mission. Understanding long-term behavior of DOE's inground contaminants will also facilitate definition of situations in which monitored natural attenuation can be considered as an appropriate remediation alternative. EM can leverage substantial work in the area of subsurface contaminant behavior done by other organizations. However, adaptation and expansion of new knowledge, technologies, and approaches to address the specific context of EM's subsurface contaminants and site conditions are needed.

Advanced Sensors, Detectors, and Data Transmission Technology for Subsurface Monitoring

EM presently does extensive monitoring of air, surface water, and groundwater at ongoing operations and closed sites to determine whether contaminants are being transported and, if so, at what rate. Much of this monitoring involves the costly and time-consuming process of taking samples in the field, transporting them to a laboratory, and analyzing the samples. A better approach relevant to all the groundwater and soil technology gaps identified in Chapter 2 would be to develop sensors that could be placed in the air, surface water, or groundwater and directly measure the contaminants of interest. These results would then be transmitted to data collection and analysis computers.

Many organizations other than EM have the same or related needs.

Substantial investments are being made in radiation detectors for homeland security purposes such as monitoring cargo. The ability to create a "lab on a chip" raises the possibility of in situ sensors that have great specificity and breadth. Wireless technology for general communication purposes has made great strides in the last decade. However, in these cases there is the need to expand the capabilities of the sensors to cover the range of species relevant to DOE, adapt advanced detector technology to EM's needs, tailor and integrate the sensors and data transmission technology, and enhance their long-term reliability. Advanced sensors that can measure the isotopic composition of nuclear materials, especially fissile materials, in the presence of a high neutron or gamma-ray background would have applications to the waste processing and facility deactivation and decommissioning gaps identified in Chapter 2.

Advanced Near-Surface Engineered Barrier Systems to Control Contaminant Release to the Environment

EM has made and will continue to make extensive use of near-surface engineered barriers to reduce water infiltration and contaminant mobilization at closed waste disposal sites and partially demolished or dismantled facilities (reactors, reprocessing plants). Such barriers include multilayer caps, synthetic cocoons, vegetative covers, French drains, subsurface barrier walls, vaults, and reactive barriers. R&D needs related to these barriers were identified in gaps WP-2 and GS-3. Such barrier systems are in use at many industrial and non-EM government hazardous chemical disposal sites, and there is an established service industry to build them. However, the radionuclides that are to be contained by the engineered barriers are unique to EM and it is unlikely that other organizations will perform the R&D in this area, especially as it concerns long-term barrier performance and the design of reactive barriers. Additionally, the design of such barriers is heavily dependent on local climate and subsurface conditions, and EM and its contractors need extensive knowledge of both.

Surface Characterization of Solid Materials

Addressing a number of technology gaps will require the capability to characterize the surface of solid materials. Examples include the characteristics of recalcitrant deposits in tanks and the surface on which they reside (Gap WP-1), characterization of the surfaces of waste forms and their degradation products resulting from interactions with groundwater (gaps WP-2 and WP-6), the surfaces of soils and engineered barriers with which contaminants may interact (gaps GS-1 and GS-4), and surfaces of equipment or construction material to be characterized or decontaminated (DD-1 and DD-3). While the necessary equipment is costly, much of it exists in the national laboratories, industrial research laboratories, and academia, and is made available as a shared resource. However, EM's applications have two unique aspects. First, EM applications involve some combinations of surfaces and contaminants (e.g., radionuclides, beryllium) that are uncommon in other applications. Second, the level of contamination, especially radioactive, can necessitate that the equipment be dedicated to radioactive operations and left in a hot cell or other location that makes its use for conventional applications impractical. This basic and applied research can benefit from leveraging, but it will also require that EM target its specific needs and applications because of the difficulty and expense of working with EM-relevant materials.

OBSERVATIONS ON THE ROLE OF EM'S OFFICE OF ENGINEERING AND TECHNOLOGY

In preparing this report, the committee confined its deliberations to the technical issues raised in its task statement. Nevertheless, in the course of our information gathering it became clear that for 20 years EM's road to introducing new technologies for cleanup has been rocky. EM's technology development efforts are perceived as having achieved only limited success in providing new technologies to make cleanup "faster, cheaper, and safer" that was the mission of EM's original Office of Technology Development, established in 1989. Lack of perceived success is manifested by, and probably synergistic with, the rather precipitous decline of funding for EM technology development in the mid-1990s described in Chapter 1 of this report.

In the spirit of assisting EM, this committee—which includes some members who participated in previous National Academies' studies of EM's technology development efforts—came to some observations that may strengthen future initiatives by EM's OET to bring new technologies into the EM cleanup effort.

The committee's first finding is that the EM Engineering and Technology Roadmap can be a key tool to ensure that key personnel and infrastructure remain available to EM during the site cleanup program. A corollary to this finding is that **a key role for OET is to manage the Roadmap,** that is, to ensure that roadmapped R&D stays on track to provide new, reliable, technologies that are appropriate and timely for use by cleanup contractors. Managing the Roadmap also means maintaining it as a living document in which R&D objectives and schedules can be changed as new cleanup challenges are encountered, the scientific state of the art advances, and lessons are learned. Managing the Roadmap includes tracking new technologies from their conception to their implementation to provide an authoritative record of their success and payback on investment; or their failure at some point in their development; or as happens often in technology development, their evolution into other technologies or unplanned applications.

Technology development of course requires stable funding not only to retain capable investigators and maintain infrastructure, as discussed in Chapter 3, but also because it usually takes new technologies at least 5 years, often more, to mature, demonstrate their advantages, and show a payoff. To help ensure stable funding for the future, OET needs to evaluate, document, and publicize its previous successes and value added to the EM cleanup, and to continue to do so as new roadmapped objectives are met.

By asking this committee to help identify strategic opportunities for leveraging its R&D resources, OET recognizes the importance, and necessity, of coordinating its work with others. Beyond the specific advice provided in Chapter 4, input to the committee and members' experience lead to the following suggestions:

- OET's outreach to form partnerships can best start with other EM offices, especially those that directly manage the cleanup. A first step would be to ensure that the technology initiatives and milestones in the roadmap are in sync with the cleanup, as recommended by this committee. A cooperative effort might be to develop an overall cleanup roadmap to which OET's technology roadmap can be linked.[11] By ensuring that its technology development is directly relevant to the cleanup, OET can establish its role within EM of resolving critical cleanup knowledge and technology gaps, ensuring the maintenance of essential capabilities and infrastructure, and managing the knowledge and technology bases needed to complete the 30-year cleanup mission.
- OET will have to take the initiative to coordinate ("leverage") its work with other offices and agencies that are potential partners and funders of relevant R&D. Further, OET will have to demonstrate that it is a good partner for others according to the principles set out in Chapter 4. Shared management of partnerships—for example, the shared management between EM and SC of the EM Science Program (NRC 1997a)—might be a good approach, especially with other DOE offices.
- OET, with assistance and cooperation from other EM offices, will have to take the initiative to reach out to cleanup contractors to identify needs, time and money constraints, practical solutions, and implementation plans. Contractors are the drivers behind the implementation of new

[11] Some initial efforts along this line were the EM Baseline Environmental Management Reports issued in 1995 and 1996 (DOE 1995, 1996) and *Paths to Closure* (DOE 1998a). Most recent, although it is not a roadmap, is EM's Report to Congress pursuant to the National Defense Autorization Act for FY 2008 (DOE 2009).

technology. One committee member stated "if site contractors don't identify the research needs and have an interest in the results, then you can develop all the neat technology that you want and it will sit there, unused."

OET would probably need to assign one or more full-time staff to each of the outreach and coordination roles stated above. When they go to work every day, these staff should be working with the sites to understand problems and looking for solutions in industry, academia, or the national laboratories; convening workshops to understand issues; and locating the best authorities nationally and internationally to address the problems. They should be working with the DOE site managers and headquarters staff who oversee site operations to ensure that contractors are aware of the best technologies available and encourage them to use it. They should be working with procurement staff to make sure use of new technology is included in the contracts and that use of the new technology is included as a metric for successful completion of the contract. Staff of the Savannah River National Laboratory, EM's designated "corporate" laboratory, can assist OET headquarters staff in developing and carrying out scientific and technical initiatives.

CONCLUSION

The fiscal year 2007 House Energy and Water Development Appropriations Report directed DOE to prepare a technology roadmap that identifies technology gaps in the current DOE site cleanup program. For assistance with its roadmapping activity DOE's Office of Environmental Management turned to the National Academies, which empaneled a committee to under take the study described in this report. The committee carried out its task with the intent of assisting and strengthening EM's roadmapping efforts.

At the beginning of the study the committee understood that the Roadmap would be a living document to help plan, justify, and increase the effectiveness of EM's R&D program in support of its site cleanup mission. The committee found that the Roadmap can be an important tool for enhancing EM's R&D efforts and has recommended improvements and periodic updates of the Roadmap. We trust that this report with its findings and recommendations will be useful to DOE and to EM.

References

AEATES (AEA Technology Engineering Services, Inc.). 2005. Calcine Bins Retrieval and Transfer System: Test Report. ESI Document No. 2156-6-004. January. Mooresville, NC: AEA Technology Engineering Services, Inc.

Ahearne, J. 2002. High-level waste melter review. Presented at Waste Management 2002 Conference, Tucson, AZ. February 24-28.

Bakari, J.M., K.M. Zied, and D.W. Seward. 2007. Development of a multi-arm mobile robot for nuclear decommissioning tasks. International Journal of Advanced Robotic Systems 4:387–406.

Barnes, A.B., D.C. Iverson, B.J. Adkins, and E. Tchemitcheff. 2008. Feasibility evaluation and retrofit plan for cold crucible induction melter deployment in the Defense Waste Processing Facility at Savannah River Site. WSRC-STI-2007-00669. Presented at Waste Management 2008 Conference, Phoenix, AZ, February 24-28.

Benson, C.H., W.H. Albright, D.P. Ray, and J. Smegal. 2008. Independent Technical Review Report: Summary and Recommendations: EM Landfill Workshop, Office of Engineering and Technology (EM-20). November 21.

Bernard, T.E. 1999. Heat stress and protective clothing: An emerging approach from the United States. Annals of Occupational Hygiene 43:321–327.

Boyd, G. 2008. Achievements and lessons from EM's focus areas and Environmental Remediation Sciences Program. Presentation to National Research Council Committee on Development and Implementation of a Cleanup Technology Roadmap, Washington, DC, April 28.

Bredt, P., A.R. Felmy, P.A. Gauglitz, D. Hobbs, S. Krahn, N. Machara, M. McIlwain, B.A. Moyer, A.P. Poloski, K. Subramanian, J.D. Vienna, and B. Wilmarth. 2008. Scientific Opportunities to Reduce Risk in Nuclear Process Science. PNNL-17699. Richland, WA: Pacific Northwest National Laboratory. Available at http://www.pnl.gov/main/publications/external/technical_reports/PNNL-17699.pdf.

Catalano, J., J.P. McKinley, J.M. Zachara, S.M. Heald, S.C. Smith, and G.E. Brown, Jr. 2006. Changes in uranium speciation through a depth sequence of contaminated Hanford sediments. Environmental Science and Technology 40:2517–2524.

Christensen, C.M., C.S. Musso, and S.D. Anthony. 2004. Maximizing the returns from research. Research-Technology Management 47(4):12–18.

Clarke, J.H., M.M. MacDonell, E.D. Smith, R.J. Dunn, and W.J. Waugh. 2004. Engineered containment and control systems: Nurturing nature. Risk Analysis 24:771–779.

Cook, J.R., and J.R. Fowler. 1992. Radiological Performance Assessment for the Z-Area Saltstone Disposal Facility (U). WSRC–RP–92–1360. Aiken, SC: Westinghouse Savannah River Company.

Cosner, R.R., E.J. Hynds, A.R. Fusfeld, C.V. Loweth, C. Scouten, and R. Albright. 2007. Integrating roadmapping into technical planning. Research-Technology Management 50(6):31–48.

Dalton, M. 2008. Open innovation: Best practices in leveraging external R&D. Presentation to National Research Council Committee on Development and Implementation of a Cleanup Technology Roadmap. Washington, DC, April 28.

Davis, N. 2008. Liquid waste operations technology development. Presentation to National Research Council Committee on Development and Implementation of a Cleanup Technology Roadmap. Augusta, Georgia, January 9.

Demmer, R., D. Peterman, J. Tripp, and K. Wright. 2007. Final Report: Capture and Collection of Radioactive Contamination on Building Exteriors. INL/EXT-07-12845. Idaho Falls: Idaho National Laboratory.

Desbats, P., H. Toubon, and G. Piolain. 2004. Status of development of remote technologies applied to COGEMA spent fuel management facilities in France. Presented at American Nuclear Society 10th International Topical Meeting on Robotics and Remote Systems for Hazardous Environments, Gainesville, FL, March 28–April 1.

DOD (U.S. Department of Defense). 1999. Defense Technology Area Plan 1999 II-32, "Advanced Lightweight Protection." Available at http://www.wslfweb.org/docs/dstp2000/DTOPDF/DTOtp.pdf.

DOE (U.S. Department of Energy). 1995. Estimating the Cold War Mortgage: The 1995 Baseline Environmental Management Report. DOE/EM-0232. Washington, DC: Office of Environmental Management.

DOE. 1996. The 1996 Baseline Environmental Management Report. DOE/EM-0290. Washington, DC: Office of Environmental Management.

DOE. 1997. Linking Legacies: Connecting the Cold War Nuclear Weapons Production Processes to Their Environmental Consequences. DOE/EM-0319. Washington, DC: Office of Environmental Management.

DOE. 1998a. Accelerating Cleanup: Paths to Closure. DOE/EM-0342. Washington, DC: Office of Environmental Management.

DOE. 1998b. Innovative Technology Summary Report: NuFab™ Anti-Contamination Suit, Deactivation and Decommissioning Focus Area. DOE/EM-0354. Washington, DC: Office of Environmental Management.

DOE. 2000. Deactivation and Decommissioning Focus Area Performance Plan. Washington, DC: DOE.

DOE. 2001. Summary Data on the Radioactive Waste, Spent Nuclear Fuel, and Contaminated Media Managed by the U.S. Department of Energy. Washington, DC: DOE.

DOE. 2002. A Review of the Environmental Management Program. Presented to the Assistant Secretary for Environmental Management by the Top-to-Bottom Review Team, Washington, DC: DOE.

DOE. 2004a. Evaluation of Amendments for Mending the ISRM Barrier, Technical Assistance Project #33, Final Technical Solutions Report: Washington, DC: Office of Environmental Management.

DOE. 2004b. Mending the In Situ Redox Manipulation Barrier, Technical Assistance Project #28, Final Technical Solutions Report. Washington, DC: Office of Environmental Management.

DOE. 2004c. Audit Report: Groundwater Remediation Activities at Hanford. DOE/IG-0655. Washington, DC: Office of Inspector General and Office of Audit Services

DOE. 2007a. Engineering and Technology Roadmap: Reducing the Uncertainty in the EM Program. Draft. Washington, DC: DOE.

DOE. 2007b. Vision 2020: The Lighting Technology Roadmap. Washington, DC: Office of Building Technology, State and Community Programs, Energy Efficiency and Renewable Energy. Available at: http://www.eere.energy.gov/buildings/info/documents/pdfs/lighting_roadmap_compressed.pdf.

DOE. 2008a. Agency Financial Report. DOE/CF-0032. November 17. Available at http://www.cfo.doe.gov/cf1 2/2008parpilot.htm.

DOE. 2008b. Engineering and Technology Roadmap: Reducing the Uncertainty in the EM Program. Washington, DC: DOE.

DOE. 2009. Report to Congress: Status of Environmental Management Initiatives to Accelerate the Reduction in Environmental Risks and Challenges Posed by the Legacy of the Cold War. Available at http://www.em.doe.gov/pdfs/NDAA%20Report-(01-15-09)a.pdf.

EPA (U.S. Environmental Protection Agency). 2005. Workshop on Decontamination, Cleanup, and Associated Issues for Sites Contaminated with Chemical, Biological, or Radiological Materials. EPA 600/R-05/083, pp. 41–46. Available at http://www.epa.gov/NHSRC/pubs/600r05083.pdf.

EPA. 2006a. Report on the 2006 Workshop on Decontamination, Cleanup, and Associated Issues for Sites Contaminated with Chemical, Biological, or Radiological Materials, pp. 61–65. Available at http://www.epa.gov/NHSRC/pubs/600r06121.pdf.

EPA. 2006b. The Technology Reference Guide for Radiologically Contaminated Surfaces. EPA-402-R-06-003. Washington, DC: Office of Air and Radiation.

EPA. 2008. Site Characterization to Support Use of Monitored Natural Attenuation for Remediation of Inorganic Contaminants in Ground Water. EPA/600/R-08/114. Cincinnati: National Risk Management Research Laboratory.

Fink, S. 2007. Tailoring inorganic sorbents for strontium and actinides. Science at Work 1(1):12–17.

Foley, T., R.G. Brown, E. Burrell, D. Del Rosso, K. Krishen, H. Moffitt, E. Orndoff, B. Santos, M. Butzer, R. Dasgupta, M. Jones, F. Herrera, and G. Vincent. 1999. Terrestrial EVA Suit = Firefighters Protective Clothing. Report 1999-01-1964. National Aeronautics and Space Administration. Available at http://ntrs.nasa.gov/archive/nasa/casi.ntrs.nasa.gov/20000081720_2000118148.pdf.

Freese, K. 2008. Successful partnering in the federal environment. Presentation to the National Research Council Committee on Development and Implementation of a Cleanup Technology Roadmap, Washington, DC, April 28.

GAO (U.S. Government Accountability Office). 2006. Nuclear Waste: DOE's Efforts to Protect the Columbia River from Contamination Could Be Further Strengthened. GAO-06-1018. Availabile at http://www.gao.gov/new.items/d061018.pdf

Gee, G.W., M. Oostrom, M.D. Freshley, M.L. Rockhold, and J.M. Zachara. 2007. Hanford Site Vadose Zone Studies: An Overview. Vadose Zone Journal 6:899–905.

Griffith, A. 2008. Opportunities for leveraging resources with the Office of Nuclear Energy. Presentation to the National Research Council Committee on Development and Implementation of a Cleanup Technology Roadmap. Washington, DC, April 28.

Hagan, W. 2008. Radiation detection R&D for homeland security. Presentation to the National Research Council Committee on Development and Implementation of a Cleanup Technology Roadmap, Washington, DC, April 28.

Hagers, J. 2007. High Level Waste (Calcine) Disposition Project. Presentation to the National Research Council Committee on Development and Implementation of a Cleanup Technology Roadmap. Idaho Falls: Idaho National Laboratory, August 29.

Halada, G. P. 2006. Complementary spectroscopic techniques to model the association of contaminant uranium with structural surfaces. Journal of Electron Spectroscopy and Related Phenomena 150:185–194.

Hazen, T.C., B. Faybishenko, and P. Jordan. 2008. Complexity of groundwater contaminants at DOE sites. Environmental Science and Technology. In review.

Honeyman, J. 2007. Hanford HLW program—Challenges to cleanup. Presentation to the National Research Council Committee on Development and Implementation of a Cleanup Technology Roadmap, Richland, WA, October 31.

Johnson, W. 2008. Interview with Bill Johnson, President, Washington River Protection Solutions, LLC. Inside Energy. October 6.

Kock, A. 2008. Leveraging research related to waste incidental to reprocessing. Presentation to the National Research Council Committee on Development and Implementation of a Cleanup Technology Roadmap, Washington, DC, April 28.

Knox, A. S., R. L. Brigmon, D. I. Kaplan, and M. H. Paller. 2008. Interactions among phosphate amendments, microbes and uranium mobility in contaminated sediments. Science of the Total Environment 395:63–71.

Kuperberg, M. 2008. The Department of Energy's Office of Science. Presentation to the National Research Council Committee on Development and Implementation of a Cleanup Technology Roadmap, Washington, DC, April 28.

Kurkumeli, A., M.I. Molokhov, O.D. Sadkovskaya, V.I. Kononov, G.B. Borisov, V.F. Savelev, and A.I. Kachurin. 1992. Use of microwave energy in reprocessing of high-level waste. Journal of Atomic Energy 73:730–734.

Kwit, C., and B. Collins. 2008. Native grasses as a management alternative on vegetated closure caps. Environmental Management 41:929–936.

LANL (Los Alamos National Laboratory). 2002. Synchrotron radiation studies aid environmental cleanup. Actinide Research Quarterly 1st qtr. Available at http://arq.lanl.gov/source/orgs/nmt/nmtdo/AQarchive/02spring/synchrotron.html.

Mauss, B. 2007. Hanford HLW program: Challenges to cleanup. Presentation to the National Research Council Committee on Development and Implementation of a Cleanup Technology Roadmap, Richland, WA, October 31.

McCracken, S. 2007. Comments to the National Research Council Committee on Development and Implementation of a Cleanup Technology Roadmap. Oak Ridge, TN, June 13.

McLaughlin, D.F., W.R. Gass, S.V. Dighe, N.D'Amico, R.L. Swensrud, and M.F. Darr. 1994. Evaluation of Plasma Melter Technology for Vitrification of High-Sodium Content Low-Level Radioactive Liquid Wastes—Demonstration Test #4 Preliminary Test Report. WHC-SD-WM-VI-014, Rev. 0. Richland, WA: Westinghouse Hanford Company.

McLaughlin, D.F., E.J. Lahoda, and M.F. Darr. 1995. Technical Information Report: Plasma Melter Operation Reliability and Maintenance Analysis. WHC-SC-WM-VI-015, Rev. 0. Richland, WA: Westinghouse Hanford Company.

Merrow, E. 2000. Problems and progress in particle processing. Chemical Innovation. (January):35–41.

Michener, T. 2007. M-3 mixing systems. Presentation to the National Research Council Committee on Development and Implementation of a Cleanup Technology Roadmap, Richland, WA, November 1.

Monken-Fernandes, H. 2008. International Atomic Energy Agency (IAEA) programs for cooperation in waste management technology. Teleconference presentation to the National Research Council Committee on Development and Implementation of a Cleanup Technology Roadmap, Washington, DC, April 28.

Moyer, B.A., J.F. Birdwell, Jr., P.V. Bonnesen, and L.H. Delmau. 2005. Use of macrocycles in nuclear-waste cleanup: A real-world application of a calixcrown in technology for the separation of cesium. Pp. 383–405 in Macrocyclic Chemistry: Current Trends and Future, K. Gloe, ed.; Dordrecht, The Netherlands: Springer

Neu, M.P., D.L. Clark, S.D. Conradson, R.J. Donohoe, J.C. Gordon, D.W. Keogh, D.E. Morris, R.D. Rogers, B.L. Scott, and C.D. Tait. 1999. Structure and stability of actinides (U, Np, Pu) under strongly alkaline radioactive waste tank conditions. Book of Abstracts, 218th American Chemical Society National Meeting, New Orleans, LA, August 22–26.

Nielsen, C., D. Gertman, D. Bruemmer, R. Hartley, and M. Walton. 2008. Evaluating Robot Technologies as Tools to Explore Radiological Environments. INL/CON-07-13568. Presented at 12th Topical Meeting on Robotics and Remote Systems for Hazardous Environments, Albuquerque, NM , March 9–12.

NRC (National Research Council). 1993. In Situ Bioremediation: When Does It Work? Washington, DC: National Academy Press.

NRC. 1994. Alternatives for Ground Water Cleanup. Washington, DC: National Academy Press.

NRC 1997a. Building an Effective Environmental Management Science Program: Final Assessment. Washington, DC: National Academy Press.

NRC. 1997b. Innovations in Ground Water and Soil Cleanup Washington, DC: National Academy Press.

NRC. 1999a. Alternative High-Level Waste Treatments at the Idaho National Engineering and Environmental Laboratory. Washington, DC: National Academy Press.

NRC. 1999b. Groundwater & Soil Cleanup: Improving Management of Persistent Contaminants. Washington, DC: National Academy Press.

NRC. 1999c. Technologies for Environmental Management: The Department of Energy's Office of Science and Technology. Washington, DC: National Academy Press.

NRC. 2000a. Alternatives for High-Level Waste Salt Processing at the Savannah River Site. Washington, DC: National Academy Press.

NRC. 2000b. Natural Attenuation for Groundwater Remediation. Washington, DC: National Academy Press.

NRC. 2000c. Research Needs in Subsurface Science. Washington, DC: National Academy Press.

NRC. 2001a. Research and Development on a Salt Processing Alternative for High-Level Waste at the Savannah River Site. Washington, DC: National Academy Press.

NRC. 2001b. Research Needs for High-Level Waste Stored in Tanks and Bins at U.S. Department of Energy Sites: Environmental Management Science Program. Washington, DC: National Academy Press.

NRC. 2001c. Research Opportunities for Deactivating and Decommissioning Department of Energy Facilities. Washington, DC: National Academy Press.

NRC. 2002. Research Opportunities for Managing the Department of Energy's Transuranic and Mixed Wastes. Washington, DC: The National Academies Press.

NRC. 2003. Improving the Scientific Basis for Managing DOE's Excess Nuclear Materials and Spent Nuclear Fuel. Washington, DC: The National Academies Press.

NRC. 2004. Contaminants in the Subsurface: Source Zone Assessment and Remediation. Washington, DC: The National Academies Press.

NRC. 2005. Improving the Characterization and Treatment of Radioactive Wastes for the Department of Energy's Accelerated Site Cleanup Program. Washington, DC: The National Academies Press.

NRC. 2006a. Improving the Regulation and Management of Low-Activity Radioactive Wastes. Washington, DC: The National Academies Press.

NRC. 2006b. Tank Waste Retrieval, Processing, and On-Site Disposal at Three Department of Energy Sites. Washington, DC: National Academies Press.

NRC. 2007a. Assessment of the Performance of Engineered Waste Containment Barriers. Washington, DC: The National Academies Press.

NRC. 2007b. Plans and Practices for Groundwater Protection at the Los Alamos National Laboratory. Final Report. Washington, DC: The National Academies Press.

NRC. 2007c. Science and Technology Needs for DOE Site Cleanup: Workshop Summary. Washington, DC: National Academies Press.

NRC. 2008. Technical and Strategic Advice for the Department of Energy, Office of Environmental Management's Development of a Cleanup Technology Roadmap: Interim Report. Letter to Mr. Mark Gilbertson, Deputy Assistant Secretary for Engineering and Technology, Office of Environmental Management, U.S. Department of Energy, Washington, DC, February 14.

Phillips, E. 2007. Groundwater remediation on the Oak Ridge Reservation. Presentation to the National Research Council Committee on Development and Implementation of a Cleanup Technology Roadmap, Oak Ridge, TN, June 13.

PNNL (Pacific Northwest National Laboratory). 2007. Building Pacific Northwest National Laboratory's Capability Replacement Laboratory, PNNL Brochure. Available at http://www.pnl.gov/rcf/pdf/crl_brochure2.pdf.

Riotte, H., and J.-G. Nokhamzon. 2008. Nuclear Energy Agency of the Organisation for Economic Co-operation and Development: Programs for cooperation in waste management. Teleconference presentation to the National Research Council Committee on Development and Implementation of a Cleanup Technology Roadmap, Washington, DC, April 29.

Roberto, J. 2008. Oak Ridge National Laboratory overview. Presentation to the National Research Council Committee on Development and Implementation of a Cleanup Technology Roadmap, Oak Ridge, TN, June 13.

Rosenberger, K.H., B.C. Rogers, and R.K. Cauthen. 2005. Saltstone Performance Objective Demonstration Document (U). CBU-PIT-2005-00146, Rev. 0. Aiken, SC: Westinghouse Savannah River Company.

Ruttan, V.W. 2001. Technology, Growth, and Development: An Induced Innovation Perspective. New York: Oxford University Press.

Scouten, C. 2008. Elements of a successful roadmap. Presentation to the National Research Council Committee on Development and Implementation of a Cleanup Technology Roadmap, Washington, DC, April 28.

Shedrow, D.M. 2008. Environmental science and biotechnology overview. Presentation to the National Research Council Committee on Development and Implementation of a Cleanup Technology Roadmap, Savannah River Site, SC, January 10.

Shuh, D., N. Edelstein, and C. Burns. 2002. Research Program to Investigate the Fundamental Chemistry of Tc. EMSP-60296. Available at http://www.osti.gov/em52/2002projsum/73778.pdf

Sink C. H., K. M. Adams, B. B. Looney, K. M. Vangelas, and N. H. Cutshall. 2004. Accelerating environmental cleanup at DOE sites: Monitored natural attenuation/enhanced attenuation—A Basis for a new paradigm. Remediation Journal 15(1):89–98.

Slowinski, G., and M.W. Sagal. 2003. The Strongest Link: Forging a Profitable and Enduring Corporate Alliance. New York: AMACOM.

Smith, B. 2008. SERDP and ESTCP: DoD's Corporate Environmental Technology Programs. Presentation to the National Research Council Committee on Development and Implementation of a Cleanup Technology Roadmap, Washington, DC, April 28.

Spears, T.J. 2008. Savannah River Site Waste Disposition Project. Presentation to the National Research Council Committee on Development and Implementation of a Cleanup Technology Roadmap, Augusta, GA, January 9.

Staubly, R., and V. Kothari. 1998. DOE EM Industry Programs Robotics Development. Report DOE/FETC/C-98/7299. Morgantown, WV: Department of Energy, Federal Energy Technology Center. Available at http://www.ntis.gov/search/product.aspx?ABBR=DE98051215

Stewart, T. 2007. Challenges to cleanup: Soils and groundwater contamination. Presentation to the National Research Council Committee on Development and Implementation of a Cleanup Technology Roadmap, Richland, WA, October 31.

Tamosaitis, W.L. 2007. Waste Treatment Plant Project: Technical accomplishments, needs, and challenges. Presentation to the National Research Council Committee on Development and Implementation of a Cleanup Technology Roadmap, Richland, WA, October 31.

Thompson, M. 2007. Hanford groundwater overview—Challenges to cleanup. Presentation to the National Research Council Committee on Development and Implementation of a Cleanup Technology Roadmap, Richland, WA, October 31.

USNRC (U.S. Nuclear Regulatory Commission). 2005. Technical Evaluation Report for Draft Waste Determination for Salt Waste Disposal. Letter Report to Charles E. Anderson, Principal Deputy Assistant Secretary, Office of Environmental Management, U.S. Department of Energy, from Larry W. Camper, Director, Division of Waste Management and Environmental Protection, Office of Nuclear Material Safety and Safeguards, December 28.

USNRC. 2008. NRC Periodic Compliance Monitoring Report for U.S. Department of Energy Non-High-Level-Waste Disposal Actions—Annual Report for Calendar Year 2007. NUREG-1911. Rockville, MD: USNRC.

Walker, J. 2008. Opportunities to leveage other R&D activities in support of Yucca Mountain. Presentation to the National Research Council Committee on Development and Implementation of a Cleanup Technology Roadmap, Washington, DC, April 28.

Ward, A. 2007. Prototype Hanford surface barrrier. Presentation to the National Research Council Committee on Development and Implementation of a Cleanup Technology Roadmap. Hanford Site Tour, Richland, WA, November 1.

Wentsel, R. 2008. Environmental Protection Agency (EPA) programs for science and technology. Presentation to the National Research Council Committee on Development and Implementation of a Cleanup Technology Roadmap. Washington, DC, April 28.

Whitaker, W. 2008. Area Completion Project: Soil and groundwater remediation and facilities deactivation and decommissioning. Presentation to the National Research Council Committee on Development and Implementation of a Cleanup Technology Roadmap, Augusta, GA, January 8.

Acronyms

ACAP	Alternative Cover Assessment Program of the EPA
ARP	actinide removal process at SRS
ASCR	Advanced Scientific Computing Research program of SC
ATR	Advanced Test Reactor at ORNL
CFD	computational fluid dynamic (model)
CSSX	caustic-side solvent extraction
D&D	deactivation and decommissioning
DDA	deliquification, dissolution, and adjustment process at SRS
DISPONET	Waste Disposal Network of the IAEA
DNAPL	dense nonaqueous phase liquid
DNDO	Domestic Nuclear Detection Office of the U.S. Department of Homeland Security
DOD	U.S. Department of Defense
DOE	U.S. Department of Energy
DWPF	Defense Waste Processing Facility at SRS
EM	DOE Office of Environmental Management
EM-20	EM Office of Engineering and Technology
EMSL	Environmental Molecular Sciences Laboratory at PNNL
EMSP	Environmental Management Science Program
ENVIRONET	IAEA Network of Centers of Excellence in Environmental Remediation

EPA	U.S. Environmental Protection Agency
ERSD	Environmental Remediation Sciences Division of OBER
ERSP	Environmental Remediation Science Program
ESTCP	Environmental Security Technology Certification Program of the DOD
ET	evapotranspiration
FCIOC	Federal Chief Information Officer Council
FRC	Field Research Center at Oak Ridge
FY	fiscal year
GNEP	DOE Global Nuclear Energy Partnership
GTCC	greater-than-Class C (low-level waste)
GWD	EM Ground Water Database
HFIR	High Flux Irradiation Reactor at ORNL
HLW	high-level (radioactive) waste
IAEA	International Atomic Energy Agency
IDN	IAEA International Decommissioning Network
INL	Idaho National Laboratory
ISRM	In Situ Redox Manipulation barrier
LANL	Los Alamos National Laboratory
LAW	low-activity (radioactive) waste
LBNL	Lawrence Berkeley National Laboratory
MCU	modular caustic-side solvent extraction unit at SRS
MNA	monitored natural attenuation
NABIR	Natural and Accelerated Bioremediation Research program
NBC	nuclear/biological/chemical (anticontamination clothing)
NE	DOE Office of Nuclear Energy
NEA	Nuclear Energy Agency of the Organisation for Economic Co-operation and Development
NPL	National Priorities List for EPA's Superfund program
NRC	National Research Council of the National Academies
OBER	SC Office of Biological and Environmental Research
OBES	SC Office of Basic Energy Sciences
OET	EM Office of Engineering and Technology (EM-20)
OR	Oak Ridge

ORNL	Oak Ridge National Laboratory
ORD	EPA Office of Research and Development
PNNL	Pacific Northwest National Laboratory
PPE	personal protective equipment
PRB	permeable reactive barrier
R&D	research and development
RCRA	Resource Conservation and Recovery Act
RI/FS	EPA Remedial Investigation/Feasibility Study
ROD	Record of Decision
RW	DOE Office of Civilian Radioactive Waste Management
SBIR	Small Business Innovation Research Program
SC	DOE Office of Science
SERDP	Strategic Environmental Research and Development Program of the DOD
SITE	Superfund Innovative Technology Evaluation program
SREL	Savannah River Ecology Laboratory
SRNL	Savannah River National Laboratory
SRS	Savannah River Site
SWPF	Salt Waste Processing Facility at SRS
TER	USNRC Technical Evaluation Report
TPB	tetraphenyl boron
TI	technical infeasibility (waiver)
TRU	transuranic, i.e., chemical elements numbered above uranium
USNRC	U.S. Nuclear Regulatory Commission
VCOC	volatile chlorinated organic compound
VOC	volatile organic compound
WTP	Waste Treatment Plant at Hanford
XANES	X-ray absorption near edge spectroscopy
XAS	X-ray absorption spectroscopy
XRF	X-ray fluorescence spectroscopy

Appendixes

Appendix A

Biographical Sketches of Committee Members

Edwin P. Przybylowicz, *Chair*, retired in 1991 after over 35 years with the Eastman Kodak Company as senior vice president and director of research. He became assistant director, Kodak Research Laboratories in 1983, was named director of research and elected as senior vice president of the company in August 1985. He has served as a commissioner of the U.S.-Polish Joint Fund for Cooperation in Science and Engineering, a program that fosters the collaboration of Polish and U.S. scientists, chairing conferences and workshops on technology transfer in Poland, the Czech Republic, and Russia. From 1994 to 1996, he was director of the Center for Imaging Science at the Rochester Institute of Technology. He is currently an elected member of the International Union of Pure and Applied Chemistry (IUPAC) Bureau and Executive Committee, and is past chair of the U.S. National Committee for IUPAC. He was elected to the National Academy of Engineering in 1990 and has served on numerous National Research Council committees, including co-chair of the Board on Chemical Sciences and Technology and chair of the Committee to Review the Worker and Public Health Activities Program. He currently chairs the Committee on Biodefense Analysis and Countermeasures for the Board of Army Science and Technology. Dr. Przybylowicz received his B.S. in chemistry from the University of Michigan and Ph.D. in analytical chemistry from Massachusetts Institute of Technology.

Allen G. Croff, *Vice Chair*, retired from Oak Ridge National Laboratory (ORNL) at the end of 2003 and is now an independent consultant. While employed at ORNL, Mr. Croff was involved in technical studies and program development focused on waste management and nuclear fuel cycles.

Mr. Croff chaired a committee of the National Council on Radiation Protection and Measurements (NCRP) that produced the 2002 report titled Risk-Based Classification of Radioactive and Hazardous Chemical Wastes; he is currently a member of the NCRP. He also chaired the Nuclear Energy Agency's Nuclear Development Committee for a decade and is currently vice chairman of the U.S. Nuclear Regulatory Commission's Advisory Committee on Nuclear Waste and a member of the Department of Energy's Nuclear Energy Research Advisory Committee. Mr. Croff has served on numerous National Research Council committees, including High-Level Waste in Tanks; Technologies for Remediation of High-Level Waste Tanks in the DOE Weapons Complex; Remediation of Buried and Tank Wastes; Long-Term Institutional Management of DOE Legacy Waste Sites; Risk-Based Approaches for Disposition of Transuranic and High-Level Radioactive Waste; and Management of Certain Radioactive Waste Streams Stored in Tanks at Three Department of Energy Sites. Mr. Croff currently serves on the Nuclear and Radiation Studies Board. He received his B.S. in chemical engineering from Michigan State University, Nuclear Engineer degree from the Massachusetts Institute of Technology, and M.B.A. from the University of Tennessee.

Richelle Allen-King is professor of geology at University at Buffalo (SUNY). She received her Ph.D. from the Department of Earth Sciences, University of Waterloo, and B.A. from the Department of Chemistry at the University of California, San Diego. Her research focuses on the geochemical processes that control the fate and transport of contaminants in groundwater and surface water. She was selected as the National Ground Water Association's Henry Darcy Distinguished Lecturer for 2003 and in that role presented her research at more than 60 national and international venues. She has served on groundwater remediation and aquifer storage committees for the National Research Council and as well as serving two terms as a member of their Water Science and Technology Board. She has also served as associate editor for the journals *Ground Water* and *Water Resources Research*. Recent funding for her research has been from the National Science Foundation, the National Institutes for Water Research, and the Department of Energy's Office of Science.

Sue B. Clark is an expert in environmental chemistry of plutonium and other actinides, chemistry of high-level radioactive waste systems, and chemistry of actinide-bearing solid phases in natural environments. She is chair of the Department of Chemistry and Westinghouse Professor of Materials Science and Engineering at Washington State University in Pullman. Previously, she was an assistant research ecologist at the University of Georgia's Savannah River Ecology Laboratory and senior scientist at Westinghouse Savannah

River Company's Savannah River Technology Center. She currently is a member of the U.S. Department of Energy's Basic Energy Sciences Advisory Committee. She received her Ph.D. in inorganic and radiochemistry from Florida State University. She has served on several National Research Council committees, including the Review of the Hanford Site's Environmental Remediation Science and Technology Plan, and she serves on the Nuclear and Radiation Studies Board.

Patricia J. Culligan is professor of civil engineering and engineering mechanics at Columbia University. Her research focuses on applying geoengineering principles to understand and control the migration of contaminants from waste disposal sites. In particular, she studies the behavior of miscible contaminants and nonaqueous-phase liquids in soil and fractured rock and the effectiveness of in situ remediation strategies for the cleanup of waste sites. She also has interest and experience in the design of land-based disposal sites for waste materials. Dr. Culligan has received numerous awards, including the Arthur C. Smith Award for Undergraduate Service (1999) and the National Science Foundation CAREER Award (1999). She is also the author or coauthor of more than 50 journal articles, book chapters, and refereed conference papers. Dr. Culligan has a Ph.D. in civil engineering from Cambridge University, England. She has served on three National Research Council committees, including Long-Term Institutional Management of DOE Legacy Waste Sites; Opportunities for Accelerating Characterization and Treatment of Waste at DOE Nuclear Weapons Sites; and Management of Certain Radioactive Waste Streams Stored in Tanks at Three Department of Energy Sites.

Rachel J. Detwiler is associate and senior engineer at Braun Intertec Corporation in Minneapolis, Minnesota. Her areas of expertise are construction troubleshooting, concrete durability, transport properties, microstructure, and test methods for concrete and cement-based materials. Dr. Detwiler previously worked as a principal engineer at Construction Technology Laboratories; an assistant professor at the University of Toronto; a postdoctoral research fellow at Norges Tekniske Hogskole, Trondheim, Norway; and a design and materials engineer with ABAM Engineers, Inc. She is a member of the American Society for Testing and Materials and a fellow of the American Concrete Institute, where she served as chair of Committee 227 on Radioactive and Hazardous Waste Management, and as a member of Committee 201 on Durability of Concrete and on the Publications Committee. She is chair of Committee 234 on Silica Fume in Concrete. She also served in an advisory role until 1986 for the initial development of a formulation of grout for the stabilization of radioactive and hazardous waste in underground storage tanks at the Savannah River Site. Dr.

Detwiler has published over 50 technical papers related to concrete microscopy, durability, and testing. She has served on two National Research Council committees, including Long-Term Research Needs for Deactivation and Decommissioning at Department of Energy Sites and Management of Certain Radioactive Waste Streams Stored in Tanks at Three Department of Energy Sites.

Thomas F. Gesell is an authority in health physics and environmental radiation monitoring. He is professor of health physics and director of the Environmental Monitoring Program at Idaho State University. Previously, he worked for the DOE Idaho Operations Office as deputy assistant manager for nuclear programs and director of the Idaho National Engineering and Environmental Laboratory's Radiological and Environmental Sciences Laboratory. Dr. Gesell was a faculty member of the University of Texas School of Public Health in Houston for 10 years. He is a fellow of the Health Physics Society and a former member of its Board of Directors. He is currently a vice president and member of the Board of Directors of the National Council on Radiation Protection and Measurements. He served a 6-year term as a member of the U.S. Environmental Protection Agency Science Advisory Board's Radiation Advisory Committee and several years as a consultant. He was a consultant to the President's Commission on the Accident at Three Mile Island. Dr. Gesell received his B.S. in physics from San Diego State University and his M.S. and Ph.D. degrees in physics (with specialization in health physics) from the University of Tennessee. Dr. Gesell has served on several National Research Council committees, including Opportunities for Accelerating Characterization and Treatment of Waste at DOE Nuclear Weapons Sites.

Gary P. Halada is associate professor of materials science and engineering and associate director of the Laboratory for Surface Analysis and Corrosion Science at the State University of New York (SUNY) at Stony Brook. He is also co-principal investigator of the joint Brookhaven National Laboratory–Stony Brook Center for Environmental Molecular Science, which is jointly supported by the National Science Foundation and the Department of Energy to study the sequestration, fate, and transport of metals in the environment. Dr. Halada's primary research focus is on surface chemistry of environment–materials interactions. His current studies focus on the association of uranium and transuranics with organic ligands and with large biomacromoleules including cellulose and cellulosic breakdown products. He has published more than 80 peer-reviewed journal and proceedings articles and several chapters on environment–materials interactions, surface chemistry and engineering, remediation, and development of novel analytical techniques. Dr. Halada received his B.S. in physics and Ph.D. in materials science, both from SUNY at Stony Brook.

Carolyn L. Huntoon is recognized for improving management practices and technical approaches to DOE site cleanup problems as the former DOE assistant secretary for Environmental Management. She held this Senate-confirmed position from July 1999 until until her retirement from the federal government in July 2001. She is currently an independent consultant in the fields of energy and aerospace. Before moving to DOE, Dr. Huntoon served in various scientific and management positions at the National Aeronautics and Space Administration (NASA), including director of the Johnson Space Center in Houston, Texas, and special assistant to the administrator of NASA in Washington, D.C. In addition, she served as an executive in residence in the George Washington University Project Management Program and spent 2 years at the White House in the Office of Science and Technology Policy. She is a fellow of the American Astronautical Society, the American Institute of Aeronautics and Astronautics, and the Aerospace Medical Association. Dr. Huntoon has been awarded the Secretary of Energy's Gold Medal, and the Outstanding Leadership, Exceptional Service, Scientific Achievement, and Distinguished Service Medals from NASA. Dr. Huntoon received her undergraduate degree from Northwestern State College, Natchitoches, Louisiana, and her M.S. and Ph.D. degrees from Baylor College of Medicine, Houston, Texas. Dr. Huntoon has authored or coauthored over 200 technical publications and books. She has served on one National Research Council committee, Opportunities for Accelerating Characterization and Treatment of Waste at DOE Nuclear Weapons Sites.

Edward Lahoda is a consulting engineer at the Westinghouse Electric Science and Technology Department. He has more than 32 years of experience in process analysis, development, design, and field support. He has extensive background in the manufacture of uranium-based fuels and operation of waste treatment and other ancillary systems. In the environmental area he was responsible for the technical development and field start-up of the Westinghouse soil washing and high-temperature thermal desorption technologies. He has chemical process design experience in processing chemical warfare agents, nuclear fuels, and high- and low-level nuclear wastes and in plasma processing of wastes and plasma production of specialty materials. He has served on committees at the Savannah River Site addressing overall operation and test data validity of the Defense Waste Processing Facility, chaired the In-Tank Precipitation Chemistry Review Panel, and was a member of the In-Tank Precipitation Replacement Review Panel. He has also participated on the committees to review the development and design of the Waste Treatment Plant and the bulk vitrification facility at Hanford. He is a member of the American Institute of Chemical Engineers. Dr. Lahoda received his B.S., M.S., and Ph.D. degrees in chemical engineering and his M.B.A. from the University of Pittsburgh. He served as a technical expert for the National Research Council Committee on Alternative High-Level Waste Treatment at the Idaho National Engineering and Environmental Laboratory and as a member of the Committee on Long-Term Research Needs for Radioactive High-Level Waste at Department of Energy Sites.

Robin Rogers is an expert in separations chemistry and does research on prevention or chemical treatment of waste streams. He holds appointments at both the University of Alabama where he is the Robert Ramsay Chair of Chemistry, Distinguished Research Professor, and director of the Center for Green Manufacturing; and at the Queen's University of Belfast (UK) where he is chair in Green Chemistry and director of the Queen's University Ionic Liquid Laboratory. Dr. Rogers' research interests include green/sustainable separation science and technology, aqueous biphasic systems, room temperature ionic liquids, environmentally benign polymer resins, crystal engineering, and radiochemistry. Dr. Rogers is the editor of the American Chemical Society journal *Crystal Growth & Design*. Dr. Rogers received his B.S. and Ph.D. degrees, both in chemistry, from the University of Alabama. He reached the rank of presidential research professor at Northern Illinois University before returning to Alabama. He served on the National Research Council Committee on Long-Term Research Needs for Radioactive High-Level Waste at Department of Energy Sites and the Committee on Risk-Based Approaches for Disposition of Transuranic and High-Level Radioactive Waste.

Gary S. Sayler is the Beaman Distinguished Professor in the Department of Microbiology, Ecology and Evolutionary Biology at the University of Tennessee, Knoxville; director of the University of Tennessee–Oak Ridge National Laboratory Joint Institute for Biological Sciences; and adjunct professor at Gwangju Institute for Science and Technology, South Korea. His research interests include microbiology; genetic engineering; molecular biology in biodegradation, bioremediation, and bioprocessing; polychlorinated biphenyls (PCBs); polycyclic aromatic hydrocarbons (PAH) in contaminated soils, sediments, and water; molecular ecology in biological waste treatment; PCR-gene probes; biosensors for bioavailable pollutants including endocrine disruptors; and nanotechnology and carbon nanofibers in microbial biofilms. Dr. Sayler has edited five books and contributed 285 publications in broad areas of molecular biology, environmental microbiology, biodegradation of PCB, PAH, BTEX, and trichloroethylene, and biotechnology. He holds 12 patents on environmental gene probing, genetic engineering for bioremediation, biosensor technology, and environmental gene expression. He received the National Institute of Health Scences' Research Career Development Award (1980-1985); was named a Top 100 Innovator in Science by *Science Digest* (1985); received the

American Society for Microbiology's Procter and Gamble Award in Applied and Environmental Microbiology (1994), the Distinguished Alumni Award of the University of Idaho (1995), and the DOW Chemical Foundation SPHERE Award (1998-2000). He was elected to the American Academy of Microbiology in 1991 and is a lifetime member. Dr. Sayler served as a member of the National Research Council Committee on Research Opportunities for Deactivating and Decommissioning Department of Energy Facilities, and on a review subcommittee on standoff explosives detection. He is currently a member of the Environmental Protection Agency's Science Advisory Board Drinking Water Committee and is an executive committee member of the Board of Scientific Counselors for EPA's Office of Research and Development.

Andrew M. Sessler is distinguished emeritus scientist at E.O. Lawrence Berkeley National Laboratory, University of California. He served as director of Lawrence Berkeley National Laboratory from 1973 to 1980 and as distinguished senior scientist from 1980 to 2001. His areas of expertise are in particle accelerator physics and plasma physics. He served as president of the American Physical Society (APS) in 1998, and past president in 1999. He has served as council member, vice chairman, and chairman of the Federation of American Scientists. He is a fellow of the American Physical Society, the American Association for the Advancement of Science, and the New York Academy of Sciences. He cofounded the human rights group, Scientists for Sakharov, Orlov, and Sharansky for which he received the first APS Nicholson Medal for Humanitarian Service in 1994. His awards include the Ernesto Lawrence Award, the U.S. Particle Accelerator School Prize, and the Wilson Prize. He received his M.S. and Ph.D. degrees in theoretical physics from Columbia University. He was elected to the National Academy of Sciences in 1990, and he served on the National Research Council's Nuclear and Radiation Studies Board.

Leslie Smith is the Cominco Chair in Minerals and the Environment at the University of British Columbia in Vancouver. His expertise is in the areas of subsurface hydrology and contaminant transport processes. His current research interests include hydrologic processes in unsaturated waste rock piles, submarine groundwater discharge to the near-shore marine environment, surface water–groundwater interactions, transport processes in fractured rock masses, hydrogeological decision analysis and risk assessment. In recent years, Dr. Smith has served on six National Research Council Committees, including the Review of the Hanford Site's Environmental Remediation Science and Technology Plan and Management of Certain Radioactive Waste Streams Stored in Tanks at Three Department of Energy Sites.

Appendix B

Presentations to the Committee

OAK RIDGE, TENNESSEE, JUNE 13-15, 2007

Welcome, *Gerald Boyd, Manager, DOE Oak Ridge Operations Office; Steve McCracken, Assistant Manager for Environmental Management (EM)*

Oak Ridge National Laboratory (ORNL) Overview, *Jim Roberto, ORNL, Deputy Director for Science and Technology*

ORNL Nuclear Facility History, *Gordon Michaels, Chief Technology Officer, Energy and Engineering Sciences*

EM Status/Central Campus Proposal, *Dirk Van Hoesen, Manager, Environmental Management Programs*

Exploratory Visualization Environment for Research in Science and Technology (EVEREST) Demonstration

Infrastructure/Capabilities Discussions, *Mark Noakes, R&D staff member in the Robotics and Energetic Systems group of the ORNL Engineering Science and Technology Division*

Wrap-Up and Closing Discussions, *Dana Christensen, ORNL Associate Laboratory Director of the Energy and Engineering Sciences Directorate*

Tour of Y-12 National Security Complex:
- Tour 1: Y-12 D&D Plans and Challenges, *David Adler, Oak Ridge Office Environmental Program*
- Tour 2: Y-12 Subsurface Plans and Challenges, *Elizabeth Phillips, Technology Development Program Manager*

Oak Ridge EM Science and Technology Plans and Challenges, *Steve McCracken, Assistant Manager, EM*

ORNL Tour and Briefing, EM Sites:
- Tour 1: Central Campus D&D Plans and Challenges and TRU Processing Facility, *David Adler, Oak Ridge Office Environmental Program*
- Tour 2: Melton Valley Caps and Long-Term Monitoring, Bethel Valley Plumes, Core Hole 8, *Elizabeth Phillips, Technology Development Program Manager*

East Tennessee Technology Park (ETTP) Site Tour:
- Tour 1: ETTP Plans and Challenges (K-25, K-27, balance of plant), *David Adler*
- Tour 2: ETTP Plans and Challenges (Statewide Record of Decision, Burial Grounds), *Elizabeth Phillips*

Questions and Answers, *Elizabeth Phillips and David Adler*

Guest Speaker at Dinner, *Gerald Boyd, Manager, DOE Oak Ridge Operations Office*

Open Session on June 15
EM's Vision for the Technology Roadmap and How the Committee Can Help, *Mark Gilbertson, Deputy Assistant Secretary, Office of Engineering and Technology, DOE-EM*

IDAHO FALLS, IDAHO, AUGUST 27-29, 2007

Welcome, *Bill Leake, Assistant Manager for Contract & Government Furnished Services and Instructions (GFSI) Delivery*

Attendees Introduction, *Jim Cooper, Team Lead, Remediation & Facilities Disposition Project*

Idaho Cleanup Project Overview Presentation, *Bill Leake, Assistant Manager for Contract and GFSI Delivery*

Idaho Nuclear Technology & Engineering Center (INTEC) Tank Farm Tank Closure Presentation and Video, *Keith Lockie, Tank Farm Project Manager*

INTEC Calcine Storage and Plans Presentation, *Jan Hagers, Calcine Project Manager*

Ground Water Remediation Actions Presentation, *Mark Shaw, Ground Water Remediation Project Manager*

Test Area North (TAN)-607 Hot Shop Demolition Presentation, *Mark Shaw, Ground Water Remediation Project Manager*

Spent Nuclear Fuel/Nuclear Materials (SNF/NM) Storage and Disposition Presentation, *Katie Hain, Team Lead, Materials Disposition Project*

Meeting Wrap-Up and Discussion Tuesday's Events, *Jim Cooper, Team Lead, Remediation & Facilities Disposition Project*

Bus En Route to Site Areas with Presentations, *Mark Arenaz, Federal Project Director, Waste Area Group-7:*
— Subsurface Disposal Area
— Accelerated Retrieval Project
— Vacuum Extraction System
— Waste Area Group-7 (WAG-7) Remediation

Tour Radioactive Waste Management Complex (RWMC):
— Accelerated Retrieval Project
— Advanced Mixed Waste Treatment Project

Sodium-Bearing Waste Project, *Bill Owca, Project Manager, Advanced Fuel Cycle R&D Support*

Idaho Nuclear Technology & Engineering Center (INTEC) Tour:
— CPP-666 (Flourinel Dissolution Process Facility)
— CPP-603 (Fuel Receiving & Storage Facility)
— CPP-659 (New Waste Calcine Facility)
— Tank Farm Area
— Integrated Waste Treatment Unit (IWTU)

TRU Waste Characteristics and Processing, *Alan Jines, Project Manager, Remote Handled TRU Waste Disposition Project*

Nuclear Energy (NE) Facilities Transfer/Beryllium Stabilization and Disposition Presentation, *Ron Gill, Project Manager, Reactor Technology Complex*

INL Capabilities and Personnel, *Linda McCoy, Lead Physical Scientist; Mike Connolly, Manager, Energy & Environment*

Tour Idaho National Laboratory Facilities

RICHLAND, WASHINGTON, OCTOBER 31-NOVEMBER 2, 2007

Welcome and Introductions, *David Brockman, Manager, DOE Richland Operations Office (RL); Shirley Olinger, Acting Manager, DOE Office of River Protection (ORP); Mike Weis, Manager, DOE Pacific Northwest Site Office (PNSO); Mike Kluse, Director, Pacific Norwest National Laboratory (PNNL)*

Hanford Operations Overview, *Roy Gephart, PNNL (site layout, production-era role); John Morse, DOE (cleanup activities, work remaining)*

PNNL Overview (traditional support of Hanford operations, expertise, facilities), *Mike Davis, PNNL; Terry Walton, PNNL*

Hanford HLW Program—Challenges to Cleanup (tank waste retrieval, analyses, processing), *Jim Honeyman, CH2M HILL; Rick Brouns/Walt Tamosaitis, Bechtel, Waste Treatment Plant (WTP); Tom Brouns, PNNL*

Groundwater Program—Challenges to Cleanup (hydrology overview, contaminant plumes, Vadose Zone Project, modeling, remediation), *Mike Thompson, DOE-RL; Bruce Ford, Fluor Hanford (FH); Terri Stewart, PNNL*

Facility D&D—Challenges to Cleanup (reprocessing canyons, reactors, Pu Finishing Plant, K-Basins), *Andy Schmidt, PNNL*

What Are Hanford's Long-term Cleanup Challenges? *Invited presentations from regulators and citizens*

Bus En Route to Site Areas with Presentations:
— PNNL campus drive-by
— HLW Tank Mock-up stop and visit

— 300 Area drive-by; on-bus presentation on Integrated Field Research Center, D&D, future use, hot cells
— 618-10, 11 burial ground drive around; on-bus presentation on problems of high-level TRU waste retrieval near operating commercial nuclear power plant

Waste Treatment Plant, *Rick Brouns/Walt Tamosaitis, Bechtel-WTP*
— Hanford Cap stop
— Tank Farm overview
— Waste Encapsulation Facility Sr/Cs capsules drive-by with on-bus briefing
— TRU retrieval areas drive-by
— Resource Conservation and Recovery Act waste disposal facility drive-by

Supplemental Tank Waste Processing
— Bulk Vitrification, *Larry Bagaasen, PNNL*
— Fractional Crystallization, DOE-ORP and CH2M HILL

Group 1: Waste Processing and Facility D&D

Facility D&D
— Enter 200-W canyon with viewing station
— Plutonium Finishing Plant drive-around and overlook

Travel to PNNL (300 Area)
En-route briefing on reactor D&D and cocooning

Radiochemical Processing Laboratory (RPL)
— Alumina Dissolution and Filtration
— Pipeline Plugging
— Ion Exchange Resins

Travel to Applied Process Engineering Laboratory (APEL)
— Drive by 336 building, large-scale mixing testing
— Pulse Jet Mixers
— Antifoam Agents and Gas Retention

Travel to Process Development Laboratory West (PDL-W)
— Pretreatment Engineering Platform
— Glass Waste Form Optimization (poster)

Travel to Environmental Molecular Sciences Laboratory (EMSL)

Group 2: Subsurface, Soil, and Groundwater

Soil and Groundwater Remediation
— CCl_4 plume remediation (200 W)
— T-Farm Pump and Treat, Surface Barrier (200 W)
— Gravel Pit that shows subsurface layers, presentation on Lysimeter facility

En-route briefings on Cr plume remediation and Sr plume remediation

Radiochemical Processing Laboratory (RPL)
— Conceptual Models: Addressing Uncertainty and Incorporating Complexity
— Translating Science to Technical Solution for In Situ Treatment of U in 300 Area
— Tank Farm Vadose Zone Sample Characterization and Tc-99 Roadmap Project

Travel to 331 past Integrated Subsurface Field Research Challenge (IFC) Site
— IFC Goals and Objectives
— Life-Cycle Monitoring
— Aquatics and Mesocosm Labs

Two groups reunite at Environmental Molecular Sciences Laboratory (EMSL)
— Science to Solution, Cs Migration—*John Zachara, PNNL*
— Science to Solution, Underpinnings of Waste Chemistry—*Andy Felmy, PNNL*

Open Session on November 2

Department of Energy, Office of Environmental Management's (DOE-EM) Needs for the Committee's Interim and Final Reports: Content, Timing, and Impacts, *Mark Gilbertson, Deputy Assistant Secretary, Office of Engineering and Technology, DOE-EM*

Science and Technology Challenges for Deep Groundwater Monitoring: Lessons Learned from the National Academies' Los Alamos National Laboratory (LANL) Groundwater Committee, *Chris Murray, PNNL; Tony Knepp, YAHSGS*

Science and technology roundtable

AUGUSTA, GEORGIA, JANUARY 8-10, 2008

Savannah River Site (SRS) Overview, *Jeffrey Allison, Manager, Department of Energy, Savannah River Operations Office (DOE-SR); Bill Spader, Deputy Manager for Cleanup, DOE-SR*

Liquid Waste Disposition Overview: Challenges and Issues, *Terry Spears, Assistant Manager for Waste Disposition Project, DOE-SR*

Nuclear Materials Stabilization Overview: Challenges and Issues, *Pat McGuire, Assistant Manager for Nuclear Material Stabilization Project, DOE-SR*

Area Completion Project Environmental Restoration/Deactivation and Decommissioning (ER/D&D) Overview: Challenges and Issues, *Wade Whitaker, Acting Assistant Manager for Closure Project, DOE-SR*

Comments from regulators and other stakeholders

SRS Tour Preparation and Discussion, *Randy Clendenning, Senior Technical Safety Analyst, DOE-SR; John Marra, Associate Laboratory Director, Environmental and Chemical Process Technology, Savannah River National Laboratory, Washington Savannah River Company (WSRC)*

SRNL Overview & Capabilities to Support EM, *John Marra, Associate Laboratory Director, WSRC*

SRNL Environmental Sciences & Biotechnology, *Debra Moore-Shedrow, Director, Environmental Science and Biotechnology, Savannah River National Laboratory, WSRC*

Tour of Aiken County Technical Laboratory Facilities

General Driving Tour
- A-Area: Savannah River National Laboratory, Savannah River Ecology Laboratory
- M-Area: Drive through M Area; Discuss D&D Progress
- Highlight Dynamic Underground Stripping (DUS) Technology & Impact, *Wade Whitaker, Acting Assistant Manager for Closure Project*
- F-Area: F-Canyon D&D Activities

F-Area: Tank Farm & Mixed Oxide Fuel Fabrication Facility (MOX) Construction Site
E-Area: Low-Level Waste Disposal Facilities Phytoremediation Pond, *Wade Whitaker, Acting Assistant Manager for Closure Project*
H-Area: H-Canyon/HB-Line, Tritium Facilities/Tritium Extraction Facility
H-Area: H-Tank Farm, Actinide Removal Process Modular Caustic Side Solvent Extraction, *Pat Suggs, Technical Development Lead*

Technology Development in the SRS Liquid Waste Mission, *Neil Davis, Project Manager, WSRC*

Defense Waste Processing Facility (DWPF) Overview, *Steve Wilkerson, Manager, WSRC*

Continue General Driving Tour
J-Area: Salt Waste Processing Facility Site
Z-Area: Saltstone Facility
P-Area: Operable Unit
Ray Hannah, Project Manager for Area Completion Project
N-Area: Chemical Metal Pesticide (CMP) Pits
Karen Adams, Project Manager for Area Completion Project

Open session on January 10

Review of the DOE-EM Engineering & Technology Roadmap and Multi-Year Program Plan, *Mark Gilbertson, Deputy Assistant Secretary, Office of Engineering and Technology, DOE-EM*

Leveraging Opportunities with the Global Nuclear Energy Partnership (GNEP), *John Marra, WSRC*

Science and technology roundtable

WASHINGTON, DC, APRIL 28-30, 2008

Welcome, *Mark Gilbertson, Assistant Secretary for Engineering and Technology, DOE-EM*

Elements of Successful Partnering, *Michael Dalton, Guided Innovation, LLC*

Successful Partnering in the Federal Environment, *Kenneth Freese, Technology Transfer Division, Los Alamos National Laboratory*

Federal Partnering with the Private Sector, *Manuel Gonzalez, Chevron Energy Technology Company*

Federal Partnerships Under the Small Business Innovation Research (SBIR) Program, *Charles Wessner, Director of the National Research Council's program on Technology, Innovation, and Entrepreneurship, National Research Council of the National Academies*

Achievements and Lessons from EM's Focus Areas and the Environmental Management Science Program (EMSP), *Gerald Boyd, Manager, DOE Oak Ridge Operations Office*

Programs and Resources for Leveraging EM Engineering and Technology Development, *Mark Gilbertson, Deputy Assistant Secretary, Office of Engineering and Technology, DOE-EM*

Panel Discussion on the Attributes of Successful Leveraging, *Edwin Przybylowicz, Committee Chair, Session Moderator*

Environmental Protection Agency (EPA) Programs for Science and Technology, *Randall Wentsel, EPA Office of Research and Development*

U.S. Nuclear Regulatory Commission (USNRC) Programs for Science and Technology, *Andrea Kock, Chief, Performance Assessment Branch, USNRC*

Research Coordination in the Department of Energy Office of Science (DOE-SC), *Mike Kuperberg, Biological and Environmental Research–Environmental Remediation Sciences Division (BER-ERSD)*

Leveraged Programs for Site Remediation, *Terry Hazen, Lawrence Berkeley National Laboratory (LBNL)*

Panel Discussion: Leveraging Core Capabilities and Infrastructure in the National Laboratories, *Phil McGinnis, ORNL; Terry Walton, PNNL; John Marra, SRNL; Terry Hazen, LBNL*

International Atomic Energy Agency (IAEA) Programs for Cooperation in Waste Management Technology, *Horst Monken-Fernandes, Waste Management Technology, IAEA; via videoconference*

Nuclear Energy Agency (NEA) Work and Perspective of the OECD/NEA, *Hans Riotte, Radiation Protection and Radioactive Waste Management, NEA;* Co-operative Programme on Decommissioning Projects, *Jean-Guy Nokhamzon, NEA; via videoconference*

The Strategic Environmental Research and Development Program (SERDP), *Brad Smith, Executive Director, SERDP*

Opportunities for Leveraging Resources with the DOE Office of Nuclear Energy (NE), *Andrew Griffith, Acting Director, Recycled Fuel Development, NE*

Opportunities to Leverage Other R&D Activities in Support of Yucca Mountain, *Jeffrey Walker, Engineer, Disposal Operations Office, DOE Office of Civilian Radioactive Waste Management (RW)*

Department of Homeland Security–Domestic Nuclear Detection Office (DHS-DNDO), *William Hagan, Director, Transformational and Applied Research, DHS-DNDO*

Elements of a Successful Roadmap, *Charles G. Scouten, Senior Associate, The Fusfeld Group*

Appendix C

Needs Matrix

This Appendix summarizes the committee's initial assessments of technology needs that led to its short list of technology gaps and the gap analyses presented in Chapter 2.

EM Cleanup Problem or Issue	Medium- to Long-Term Technology Need	Relevant to Site? [(x) = our view]				Technology Approaches
		OR	HAN	ID	SR	
Supporting decisions concerning "how clean is clean": tanks, subsurface remediation, facility D&D	Technical basis to support evaluation of alternative risk-informed approaches	x	x	x	x	Develop a generic technical approach and data insofar as possible for doing site-specific performance and cleanup technology assessments for capped burial grounds, capped engineered structures (tanks, vaults, canyons, reactors), and the extent of decontamination required
Increasing the rate at which low-activity waste (LAW) is converted to a final waste form	Improved/alternative LAW waste treatment technology		x		(x)	1. Improved/alternative LAW stabilization technology 2. Improved/alternative tank waste processing technology to feed LAW stabilization
Unexplained increases in mercury (Hg) concentrations in the environment	Understanding Hg in the environment	x		(x)		Develop a mechanistic understanding of the chemistry, transport, and biology of Hg in the environment and key receptor species
Ensuring continuity of knowledge/experience due to impending retirements and the inability to maintain critical and stable level of funding to support high-priority technology areas	Continuity of knowledge	x	x	x	x	Systematically identify and prioritize threats to the continuity of knowledge
Removing nonradioactive chemicals such as Al, Cr, Na from sludge	Improved separations technology		x		x	More selective and efficient approaches to removing nonradioactive chemicals from sludge wastes Develop accurate simulants
Removing key radionuclides—Cs, Sr, TRU, Tc—from sludge	Improved separations technology		x		x	More selective and efficient approaches, including alternatives, to removing key radionuclides from supernatants

Increasing the rate at which proccessed high-level waste (HLW) is converted to a final waste form	Improved HLW stabilization technology		x		x	1. Improve instrumentation to reduce batch holding times 2. Alternative high-throughput melters 3. Improve ability of glass to incorporate waste 4. Alternative HLW forms 5. Develop accurate simulants
Tank cleanout with many physical obstructions (coils, debris, and annuli)	Improved retrieval technology		x		x	Chemical approaches to removing waste from surfaces that do not degrade the tanks or cause downstream problems Physical approaches to isolate or corral debris while retrieving waste to the maximum extent practical Physical approaches to demolish and remove internal tank structures to allow access for waste retrieval and reduce intrusion pathways
Lack of capability to predict contaminant mobility in varying situations. Includes bioactivity, reactivity with environment, releases from deactivation and decommissioning (D&D), changes in hydrology with building removal	Understanding contaminant mobility	x	x	x	x	Broad-based understanding of source mobilization, transport, and degradation for organic and inorganic species in the subsurface
Predicting the long-term performance of caps, liners, and reactive barriers	Understanding the behavior of engineered containment barriers	x	x	x	x	1. Broad-based understanding of the performance of caps and liners in the context of their natural environment 2. Broad-based understanding of the performance and longevity of reactive barriers

EM Cleanup Problem or Issue	Medium- to Long-Term Technology Need	Relevant to Site? [(x) = our view]				Technology Approaches
		OR	HAN	ID	SR	
Faster methods to gather the solids from waste tanks while minimizing the amount of water introduced into tanks	Improved retrieval technology	(x)	x	x	x	Faster physical approaches (e.g., pushers) that still keep water volume increase low
Predicting the long-term performance of saltstone (including effects of chemicals from sludge washing), saltstone vaults, grouted tanks, grouted pipelines	Understanding the behavior of engineered containment barriers	(x)	x	x	x	1. Extend the current duration for predicting the performance of cementious materials not containing waste constituents 2. Obtain capability to predict the performance of cementious materials containing waste constituents
Separating solids from supernatant and mobilized sludge streams	Improved separations technology		x		x	1. More efficient, reliable, and compact technologies to separate solids from liquids 2. Develop accurant simulants
Supporting changes in regulations that allow improvement in productivity and safety for workers, for example, transite removal	Technical basis to support evaluation of alternative risk-informed approaches	x	x	x	x	1. Develop a comparison of the risks and costs of the current approach to removing transite from facilities and an approach involving use of remote systems. 2. Develop remote systems for removing transite panels from buildings well aboveground to provide a basis for the comparative evaluation
Characterization and remediation of dangerous buildings contaminated with hazardous materials	Remote techologies	x	(x)	(x)	(x)	1. Automated remote systems and robots deploying sensors for radionuclides, asbestos, and organic chemicals 2. Automated remote systems and robots deploying tools to remove equipment, decontaminate surfaces and remove deposits of radionuclides and chemicals, and remove asbestos from pipes and equipment

Nuclear materials, wastes, and spent fuels having no disposal path	Technical basis to support evaluation of alternative risk-informed approaches	x	x	x	x	Develop generic technical approaches and data for site-specific use, and specific techncial approaches and data for multisite issues, to provide a technical basis for risk-informed changes to treatment, waste acceptance criteria, and regulations
Developing cost-effective, in situ remediation of carbon tetrachloride, Cr^{+6}, Cs, U, Sr, and other contaminant plumes	Improved remediation technology	(x)	x	(x)	(x)	1. Capability to retrieve most of the toxic materials in subsurface plumes 2. Capability for fixing or destroying toxic species in subsurface plumes that is sustainable over the long term
Retrieval or control of buried contaminants (including pyrophoric material) from hot-spot sources	Improved remediation technology	x	(x)	x	(x)	1. Capability to locate hot spots 2. Capability to retrieve hot spots and backfill without impairing function of existing engineered barriers 3. Capability to stabilize hot spots in situ
Disaggregation/size reduction of residual clinkers, gravel, and recalcitant surface deposits in waste tanks	Improved retrieval technology		x		x	1. Chemical approaches that do not degrade the tanks or cause downstream problems 2. Focused physical approaches (e.g., focused water jets, grinders)
Efficient and safe human access to dangerous environnments	Better worker protection equipment	(x)	(x)	x	(x)	Lighter and cooler PPEs to allow workers to safely remain longer in the presence of hazardous airborne materials
Handling shear thickening sludges	Understanding of sludge behavior		x			1. Predictive understanding of sludge behavior (chemical and physical) as a function of its content and environment 2. Develop accurate simulants
Underground piping cleanout and closure	Improved retrieval technology	(x)	x	x	x	Technology to clean and close clogged pipelines

EM Cleanup Problem or Issue	Medium- to Long-Term Technology Need	Relevant to Site? [(x) = our view]				Technology Approaches
		OR	HAN	ID	SR	
Long-term monitoring of closed facilities and engineered barrier systems	Improved monitoring technology	x	x	x	x	Capability for sustainable remote monitoring around and within engineered facilities and barriers
Long-term monitoring strategies and devices for contaminants in soil and groundwater and deep vadose zones	Improved monitoring technology	x	x	x	x	Capability for sustainable remote subsurface monitoring
Determining moisture flux through representative waste sites, including vegetated and graveled surfaces, accounting for seasonal variations in precipitation and heating and development of barriers to moisture infiltration	Improved subsurface characterization	(x)	x	(x)	(x)	Develop real-time capability to measure water flux through engineered barriers
Predicting the long-term performance of construction concrete (e.g., canyons, reactors) that is intact or reduced to grouted rubble	Understanding the behavior of engineered containment barriers					1. Broad-based understanding of the performance of engineered barrier systems in the context of their natural environment 2. Broad-based understanding of the performance of reactive barriers 3. Extend the current duration for predicting the performance of cementitious materials not containing waste constituents 4. Obtain predictive capability to predict the performance of cementitious materials containing waste constituents

Increased HLW storage tank capacity	Technical basis to support evaluation of alternative risk-informed approaches		x		x	1. Develop a risk-informed comparison of the current approach and an approach involving additional waste tanks which would lead to more waste being retrieved by using more water 2. Extend the current understanding of tank waste chemistry to provide a basis for combining wastes now believed to be incompatible
Detailed analysis for determining proposed alternative end-state acceptance criteria or performance assessment for large contaminated structures such as canyon buildings	Technical basis to support evaluation of alternative risk-informed approaches		x		x	Develop a generic technical approach and data insofar as possible for doing site-specific performance and remediation technology assessments for greenfield and rubbleized/grouted facilities such as canyons and reactors
Determining how to characterize the subsurface if there is a large amount of subsurface piping or other structures	Improved subsurface characterization	(x)	x	(x)	(x)	1. Extended the capabilities of High Resolution Resistivity/Subsurface Geophysical Exploration to areas with large amounts of infrastructure and to areas with comingled plumes from nearby cribs and trenches 2. Develop noninvasive subsurface characterization technologies for humid sites
Field screening methods for packaged waste to characterize actinide content with high radiation backgrounds	Improved technology for characterizing difficult wastes and materials	x	x	x		Advanced, real-time assay and characterization techniques for fissile material, key radionuclides, and other paramaters (e.g., moisture content) for materials with high neutron and/or gamma background
Reducing future Al additions to HLW tanks	Remove Al cladding from fuels before dissolution				x	Technology to efficiently and inexpensively declad Al-clad fuels

EM Cleanup Problem or Issue	Medium- to Long-Term Technology Need	Relevant to Site? [(x) = our view]				Technology Approaches
		OR	HAN	ID	SR	
Technical impracticality waiver	Technical basis to support evaluation of alternative risk-informed approaches	x				Develop a generic technical approach and data insofar as possible for doing site-specific performance and remediation technology assessments to support consideration of technology impracticability waivers
Real-time detection of airborne contaminant release during demolition of Be-contaminated buildings	Improved sensors	x				Sensors capable of real-time detection of Be on surfaces and in the air
Removal of contaminants from cementitious matrices	Improved decontamination technology		x			Advanced methods to leach/migrate contaminants from cementitious matrices
Acceptability of treated Idaho sodium bearing waste in Waste Isolation Pilot Plant (WIPP)	Technical basis to support evaluation of alternative risk-informed approaches			x		Develop a risk-informed basis for disposing of sodium-bearing waste and other TRU wastes in WIPP
Predictability and control of hydrogen generation	Understanding of sludge behavior		x			Predictive understanding of sludge behavior (chemical and physical) as a function of its content and environment
Regulator-approved technologies for characterizing soil and groundwater in the field	Improved subsurface characterization	(x)	x	(x)	(x)	Develop technologies acceptable to regulators (with regulatory involvement) to characterize contaminants in the field
Disposition approach for Idaho calcine	Technical basis to support evaluation of alternative risk-informed approaches			x		Develop a risk-informed comparison of disposing of calcine, vitrified calcine, and separated and vitrified calcine in an HLW repository

Mobile waste retrieval systems for small tanks	Improved retrieval technology	x		Mobile, flexible, smaller retrieval devices that can be easily transported from tank to tank
Processing SRS Tank 48 waste	Destruction of residual tetraphenylborate (TPB) and other organic chemicals		x	Improved understanding of TPB degradation and the effects of tank contents thereon

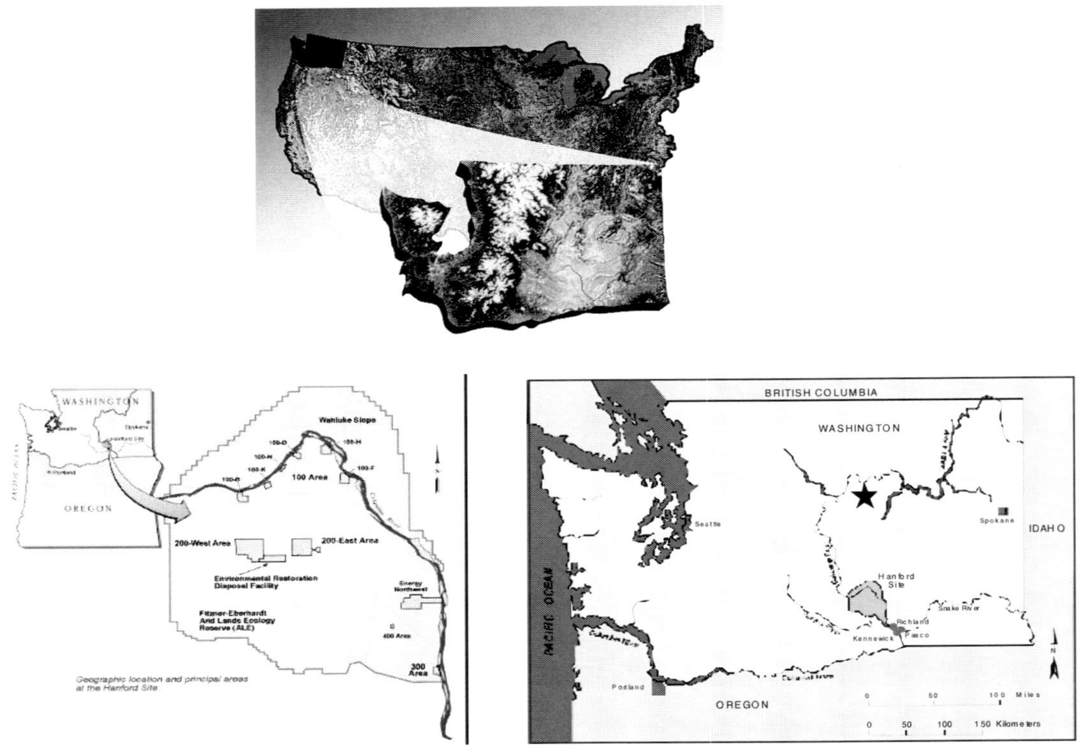

FIGURE D.1 The Hanford reservation is a 670 square mile site in southeastern Washington. Industrial-scale production of nuclear weapons materials began on the site in 1944.
SOURCE: Department of Energy.

Appendix D

Hanford Reservation

INTRODUCTION

The National Research Council Committee on Development and Implementation of a Cleanup Technology Roadmap held its fourth meeting in Richland, Washington, on October 30-31 and November 1, 2007. The purpose of the meeting was to obtain information relevant to the committee's Statement of Task (SOT) through presentations and tours by Department of Energy (DOE) staff and their contractors.[1]

This appendix provides a factual summary of the information related to the four items in the committee's SOT obtained during the meeting, the site visits, and documents provided to the committee. This appendix first describes the history and status of the DOE site at Hanford to provide perspective on the range of cleanup issues being managed by the DOE Office of Environmental Management (EM). The next sections summarize information presented to the committee, which guided the committee's deliberations in addressing its SOT as described in the main text. This appendix thus provides support for the findings and recommendations developed by the committee.

HISTORY

In March of 1943, 670 square miles in southeastern Washington State were chosen to be the site for the plutonium manufacturing operations of

[1] The agenda for this meeting is shown in Appendix B.

the Manhattan project, see Figure D.1. Designated the Hanford Reservation, the area became the first industrial-scale plutonium production site in 1944, and its role was to provide the United States with nuclear weapons material (Gephart 2003). Hanford plutonium was used in the "Fat Man" atomic bomb dropped on Nagasaki in World War II. After the war the Atomic Energy Act of 1946 removed Hanford from military control and placed it under the civilian-run Atomic Energy Commission, which later became the DOE (Gephart 2003).

During the early years, radiation exposure and waste disposal were not regarded as significant issues versus production needs and, therefore, not closely regulated. The first reactors built were called single-pass reactors because the cooling water ran only once through the core, where it became contaminated with radioactive activation and sometimes fission products, before being discharged to the Columbia River or soil. Water recirculation was included in the last of nine reactors built onsite; it came online in 1964 (Gephart 2003).

Five reprocessing plants were built in the 1940s to 1950s: T Plant, B Plant, U Plant, REDOX Plant, and Purex Plant, which generated huge volumes of waste. T and B plants produced an average of 4,000 gallons of waste liquid for each ton of spent fuel reprocessed. On average, the REDOX and Purex plants produced 4,600 gallons and 500 gallons, respectively, of liquid per ton of fuel. Table D.1 shows the amount of fuel reprocessed at each plant and their operating histories (Gephart 2003).

Depending on its source and radionuclide content, the waste was either released to the environment or pumped into large tanks. Highly radioactive waste was typically stored in tanks, and resulted in some leakage into the ground. Solid waste was buried in landfills or stored in surface facilities. Water that contained low- to intermediate-levels of radioactivity was released into the Columbia River, ponds, trenches, or cribs (Gephart 2003). In 1970, transuranic-contaminated solid waste began to be separated from low-level waste and placed in retrievable storage for subsequent offsite ship-

TABLE D.1 Uranium Fuel Processed in Hanford Reprocessing Plants

Plant	Fuel Reprocessed (tons)	Operating History
T and B	8,900 (8%)	1944-1956
REDOX	24,600 (23%)	1952-1967
PUREX	73,100 (69%)	1956-1972, 1983-1990
Total	106,600 (100%)	

SOURCE: Gephart (2003, p. 1.26)

ment to a geologic repository (Gephart 2003). Gaseous effluents from the reactors and reprocessing plants were released into the atmosphere. Nuclear material remained stored in surface facilities.

There are many unknowns about the volumes and physical, chemical, and radiological characteristics of the waste generated, stored, and released at Hanford. The wastes in the tanks are complex chemical mixtures. They have been created from multiple reprocessing techniques that include bismuth phosphate and solvent extraction using hexone or tributyl phosphate. Also, other processes have been used to recover radionuclides. Acidic waste streams were made caustic by adding concentrated sodium hydroxide, and sometimes the tanks were used as dumps for miscellaneous waste such as experimental fuel elements, ion exchange columns, and plastic bottles containing plutonium and uranium. Diatomaceous earth and cement were also added to some tanks to soak up liquids (Gephart 2003).

The Hanford tanks contained about 54 million gallons of chemical and radioactive waste. This accounted for 10 percent of the volume of waste that was originally generated. The other 90 percent was treated, released to the ground, or evaporated into the air. The high-level waste (HLW) is stored in 177 single- and double-shelled tanks. The 190 million curies of waste in these tanks make up about 50 percent of Hanford's radioactivity inventory. About 67 of the 149 single-shelled tanks have or are suspected to have leaked up to 1.5 million gallons of waste into the ground (Gephart 2003). To prevent additional releases, all drainable liquid waste in the single-shell tanks has been pumped into the newer double-shell tanks. None of the double-shell tanks have leaked, although they have long surpassed their original life expectancy (Gephart 2003). Tank cleanup is now scheduled to be completed in 2042, which means the oldest single-shelled tanks will be about a century old.[2]

Information presented to the committee shows that EM has made significant progress in cleanup and site remediation. However, removal of the majority of the waste from the tanks, solidification of the high- and low-activity portions of tank wastes, deactivation and decommissioning of many structures, and remediation or monitoring of much of the contaminated subsurface still remain to be done. Most of the site's nearly 400 million curies and 400,000 to 600,000 tons of chemicals remain to be dealt with onsite.

DOE has divided these tasks into separate programs that are administered by two DOE field offices:

---

[2] Since the briefings in 2007, DOE has issued the River Protection Project System Plan, Revision 3A which projects a tank cleanup completion date of 2049 for a reference case that approximates the key features of the current baseline and underlying technical basis (available at http://www.hanford.gov/orp/uploadfiles/ORP-11242_Rev-3A_(Released).pdf).

1. Tank cleanup (Office of River Protection, ORP), and
2. Other site cleanup and remediation activities (Richland Operations Office, RL).

The participants at the committee's Richland meeting and site visits included representatives from EM, ORP, RL, the DOE Office of Science (SC), the Pacific Northwest National Laboratory (PNNL), site cleanup contractors, regulatory agencies, Native Americans, and other citizens.

In carrying out its tank cleanup mission, the ORP will be dealing with more than 50 million gallons of tank waste in 177 underground tanks. ORP is also responsible for construction of the Waste Treatment Plant (WTP), a $12 billion project that has been described as the world's largest nuclear chemical processing plant (Tamosaitis 2007). The private company Bechtel National, Inc. is contracted to design and construct the plant and the private company CH2M HILL is contracted to operate the tank farms and assist ORP in planning and optimization of its mission. This includes leading technology roadmapping efforts and systems planning. CH2M HILL also works with EM-20's science and technology programs (Honeyman 2007).[3]

The RL mission includes the disposition or remediation of:

- 2,100 metric tons of spent nuclear fuel (SNF);
- 18 metric tons of plutonium-bearing materials, which is in various forms;
- 80 square miles of contaminated groundwater;
- 25 million cubic feet of buried or stored solid waste in 175 waste trenches;
- about 1,700 waste sites and 500 contaminated facilities, including five reprocessing canyons and nine reactor complexes; and
- 1,936 capsules of cesium and strontium containing 130 million curies of radioactivity (Morse 2007).

DOE, the U.S. Environmental Protection Agency (EPA), and the State of Washington Department of Ecology signed a comprehensive cleanup and compliance agreement on May 15, 1989. The Hanford Federal Facility Agreement and Consent Order, usually called the Tri-Party Agreement (TPA), is an agreement for achieving compliance with the Comprehensive Environmental Response, Compensation, and Liability Act remedial action provisions and with the Resource Conservation and Recovery Act treat-

[3] Since the site visit in 2007, ORP has awarded the tank operations contract to Washington River Protection Solutions, which replaced CH2M HILL on October 1, 2008.

ment, storage, and disposal regulations and corrective action provisions (DOE-ORP/RL 2007). According to the EPA's original assignment of potential hazards, Hanford has four different sites that qualified for the National Priorities List. The four areas include the 100 Area (reactors), 200 Area (fuel reprocessing), 300 Area (which includes the waste burial sites located a few miles north), and the 1100 Area (Gephart 2003).

EM manages its cleanup work according to legal provisions of the TPA. Honeyman (2007) noted that if future TPA changes reduce the amounts of contamination allowed to be left onsite after the cleanup, and therefore increase the requirements for removal of waste from tanks or from burial sites, these changes could introduce new technical challenges for accomplishing the cleanup.

CLEANUP PROGRAMS AND CHALLENGES

This section describes ongoing cleanup programs and challenges as presented to the committee during its open meeting sessions at Richland, Washington, and its visits to the Hanford site and to PNNL. The information is organized according to the program areas of the EM Science and Technology Roadmap (DOE 2008).

Roadmap Area: Waste Processing

"Hanford waste tanks are, in effect, slow chemical reactors in which an unknown but large number of chemical (and radiochemical) reactions are running simultaneously. Over time, the reaction dynamics and compositions have changed and will continue to change" (Colson et al. 1997, p. B-11).

Cleanout and closure of the Hanford tanks is a major EM challenge. As described above, tank waste is highly heterogeneous among Hanford's 177 tanks, and it is also heterogeneous within any given tank. This is due to the variety of fuel reprocessing and plutonium recovery processes used at the site, especially in the early years of production, and the fact that the acidic reprocessing waste was neutralized and made alkaline for extended storage in Hanford's carbon steel waste tanks.

As a result of neutralization, the tanks may contain one or more of the following: (1) an insoluble "sludge" that contains precipitated metals, fission products, and most actinide elements; (2) a salt cake that contains water-soluble chemicals and some fission products, notably Cs-137; and (3) a supernatant salt solution. These wastes are to be retrieved and, depending on their composition, prepared for disposal according to four methods:

- About 8 to 9 million gallons (17 percent of total) of the tank waste will be vitrified in the WTP's HLW facility,
- About 14 million gallons (26 percent of total) will be vitrified in the WTP's low-activity waste (LAW) facility,
- About 28 million gallons (52 percent of total) will be vitrified in a supplemental bulk vitrification facility, or by another supplemental process, and
- About 2 to 3 million gallons (5 percent of total) that qualify as transuranic (TRU) waste will be packaged and sent to the Waste Isolation Pilot Plant (WIPP) in New Mexico, which is DOE's TRU waste disposal facility (NRC 2006; Mauss 2007).

Waste Treatment Plant

The purpose of Hanford's WTP is to enable tank closure by processing the tanks' contents into appropriate waste forms for disposal. Constructing and operating the Hanford WTP is an unprecedented, one-of-a-kind challenge for DOE. The WTP is planned to begin full operation in 2018.

At the time of the committee's visit, the WTP was about 70 percent designed and 30 percent constructed. Brouns (2007) noted that construction had been delayed by concerns regarding the WTP's ability to withstand seismic events, and that construction resumed after these concerns were resolved by PNNL and other expert evaluations.

The WTP includes four main facilities:

1. Waste pretreatment,
2. LAW vitrification,
3. HLW vitrification, and
4. Laboratory analyses, plus a "balance of facilities" building.

Tamosaitis (2007) noted that the processing facilities rely on many first-of-a-kind technologies or applications of technologies.

Pretreatment

The purpose of waste pretreatment is to process incoming tank waste in order to obtain two waste streams for vitrification either as HLW or as LAW. A quarter-scale pretreatment engineering "platform" is being installed to test and demonstrate pretreatment operations including integrated sludge washing, leaching, and waste concentration. These tests will use nonradioactive waste simulants. The test facility does not include the ion exchange operations.

Brouns (2007) presented key WTP technology issues identified by an External Flowsheet Review Team:

- Plugging of process piping,
- Mixing vessel erosion,
- Mixing system design,
- Process operating limits, and
- Scale-up demonstration of sludge leaching and filtration.

The pretreatment engineering platform is designed to help address the last two items.

As examples, Brouns described the EFRT finding that mixing system designs may result in insufficient mixing, especially with large particles, small dense particles, and rapidly settling Newtonian slurries, and that neither caustic leaching nor the oxidative leaching process has been demonstrated at greater than bench scale. Michener (2007) described mixing requirements in the WTP as being on the forefront of mixing science. In particular, the pulse jet mixers, which are an important no-moving-parts component of the WTP, introduce unique solids lifting behavior. Tamosaitis (2007) also described current research and development (R&D) on these and similar issues, including waste stream rheology in pipes and during mixing in tanks, sludge washing, hydrogen generation, process chemistry, and online instrumentation.

In describing the pretreatment ion exchange operations, intended to remove cesium-137 and some other radionuclides from the LAW stream, Tamosaitis explained that there is need for better understanding and optimization of waste filtration and the Cs-removal ion exchange resin. The resin is an organic polymer and subject to degradation by chemicals and radiation. More resistant inorganic ion exchange materials exist but cannot be eluted (Tamosaitis 2007).

LAW Vitrification and Supplemental Treatment

The WTP will have the capacity to vitrify only about a third of the tank waste identified as LAW, as noted earlier. Several supplemental processes including those referred to as bulk vitrification and steam reforming are being developed as options to process the remaining two-thirds. Other WTP throughput issues indicate the need for supplemental LAW treatment. Examples are the handling of diverse input streams, behavior of solids, and response to process upsets (Tamosaitis 2007). Tamosaitis also listed improved waste forms, glass formulations, and melters as future technology needs for enhancing throughput. A supplemental pretreatment process, fractional crystallization, to separate radioactive cesium from the LAW salt stream is also under development.

Bagaasen (2007) described his work to resolve an important problem with bulk vitrification. In this process, waste and glass-forming material (frit) are mixed into a refractory-lined box, and the mixture is melted via

Joule heating. It appeared that the radionuclide Tc-99 would migrate from the melt into the refractory and hence not be properly incorporated into the glass. Starting with a conceptual model, Bagaasen was able to modify the frit and add carbon sources to increase the melt viscosity, and to modify the "cold cap" barrier between the melt and refractory, to resolve the problem. These results have been verified by full-scale tests using nonradioactive rhenium as a surrogate for the Tc-99 (Bagaasen 2007).

Fractional crystallization was recommended for evaluation as the result of a series of workshops held in 2002. The EM Office of Cleanup Technology (EM-21) began funding process development in 2005. The process development team includes: AREVA NC, Swenson Technologies, Savannah River National Laboratory, Georgia Institute of Technology, and CH2M HILL.

Fractional crystallization uses evaporation and crystallization to separate radioactive isotopes from the nonradioactive sodium salts that make up a large fraction of Hanford tank waste. As the water in the waste solution evaporates, nonradioactive sodium salts crystallize. Radionuclide ions like $^{137}Cs^{+}$ are too large to substitute for Na^{+} ions in the sodium salt crystal, so the radionuclides tend to remain in the liquid phase. Separation of the crystals from the remaining solution completes the process. While there are clear advantages, for example, no new chemicals have to be added to the waste, there are a number of R&D challenges at the present time.

Honeyman (2007) observed that Hanford's waste processing rates may need to increase in order to maintain the tank closure schedule. R&D could be targeted toward:

- Providing greater supplemental capacity for LAW waste treatment or making current waste treatment more efficient and rapid,
- Developing fractional crystallization as an additional or alternative pretreatment technology, and
- Reducing secondary wastes.

Honeyman concluded that bulk vitrification and fractional crystallization are potentially applicable at other DOE sites as well as Hanford.

Research Needs and Capabilities

As research needs for WTP operation, Tamosaitis (2007) included precipitation/gelation modeling and prediction, non-Newtonian computational fluid dynamics modeling, and simulant development.[4] These are needed to reduce the risks of pipe or equipment plugging in the WTP. This is especially important for WTP's "black cells" in which maintenance will not be possible. For steps in which the waste or its radionuclides are concentrated, hydrogen mitigation and handling is a challenge. Improved online instrumentation for process control and quality assurance is also needed (Tamosaitis 2007). Alternatives to borosilicate glass that can incorporate more waste per unit volume and/or be fabricated more efficiently, for example, iron-phosphate glass, might be developed through further research (Smith 2007).

Tamosaitis (2007) described national laboratory capabilities that will be needed to support construction, start-up, and operation of the WTP for the next 25-30 years:

- Radiochemistry,
- Modeling (all forms),
- Glass/waste form development,
- Hot cells,
- Analytical development and support,
- Pilot testing facilities,
- Chemical engineering and chemistry,
- Materials technology, and
- Continuity of technical knowledge.

He noted that there is a special need to ensure a future supply of technical personnel. This was highlighted by the EFRT as well as observations about the number of retiring baby boomers, decline in engineering graduates, commercial competition, and the extended schedule for WTP operation. Over the next 10 years, half of the Hanford workforce holding critical waste treatment operation and research skills will retire.

Hanford Tank Issues

The EM roadmap (DOE 2008) includes waste storage, retrieval, and tank closure within the waste processing program area. Information on these topics was presented during the committee's Richland meeting and the Hanford site visit.

---

[4] Engineering-scale process tests are usually run in facilities that cannot utilize actual radioactive waste. There is a risk that the simulated wastes may not behave the same way as actual wastes and thus provide misleading results.

Storage

Brouns (2007) suggested three technology needs to support the continued use of Hanford's double-shelled waste tanks. The first is improved understanding of tank corrosion mechanisms, especially in the vapor space above the waste and at the liquid line. The second is for new approaches that may provide increased waste storage capacity. The third is for remote inspection systems, including ultrasonic testing over large areas. Methods are also needed to assess the integrity of the single-shelled tanks, especially as waste retrieval schedules are impacted by delays in start-up of the waste processing facilities.

Retrieval

At the time of the committee's visit, seven single-shell tanks had been emptied, waste retrieval was in progress for two tanks, and two tanks were being outfitted for retrieval (Mauss 2007). Hanford's waste retrievals have met the tank cleaning requirements of the TPA, but continued improvements to make the retrievals faster and less expensive will be sought during the remaining decades of the tank cleaning work. For example, it cost $50 million to retrieve waste from the first Hanford tank (106C); that expense has now been reduced to about $10 million per tank (Honeyman 2007). Honeyman suggested improved technologies to:

- Reduce the size of residual clinkers and gravel,
- Gather the solids without introducing a lot of water,
- Disaggregate consolidated materials,
- Handle shear-thickening sludge,
- Speed up low-water retrieval methods used in leaking tanks, and
- Deal with the last few percent of waste in tanks.

Honeyman also suggested that increasing the amount of material removed from tanks would require some new technology—perhaps improved robotics, chemical dissolution of recalcitrant wastes, and fluid mechanics.

Other technology needs for waste retrieval include improving the operating life of in-tank cameras and lights, which are exposed to heat, radiation, and corrosive vapors, and improved techniques for installing openings in the top of the waste tanks (called "risers") in order to gain better access to the waste for sampling and retrieval (Brouns 2007).

Meeting needs for waste characterization for retrieval would also help provide information needed for waste processing. These include:

- Better ways to mix and sample double-shell tank waste,

- Measurements of HLW slurry hardness and abrasiveness,
- Online monitoring of the percent of solids in slurry, and
- Waste slurry transport characterization and pipeline unplugging (Brouns 2007).

Tank Closure

Brouns (2007) cited technology needs for tank closure as being:

- Postretrieval, long-term immobilization of residues,
- Pipeline characterization, and
- Demonstration of the closure and monitoring of a given waste tank area.

A draft environmental impact statement (EIS) for tank closure is expected in early 2009 with a final EIS and Record of Decision perhaps a year later. Although there has been no official word, it is anticipated that closure will proceed as it has at Idaho and the Savannah River Site, which is filling the tank with multiple layers of tailored grout, grouting pipes associated with the tank, and then covering the tank or group of tanks with a low-permeability clay cap.

Roadmap Area: Groundwater and Soil Remediation

For over five decades, liquid contaminants (~450 billion gallons) were released to the ground through injection wells, French drains, trenches, ponds, and cribs, contaminating both the vadose zone and groundwater (Thompson 2007). Two federal- and state-licensed liquid treatment plants were built in the early 1990s enabling better monitoring and control of the previously untreated discharge water (Gephart 2003). Untreated wastewater has not been discharged into the ground at Hanford since 1995.

The groundwater under about 80 square miles (15 percent) of the site is contaminated with regulated constituents at concentrations exceeding the drinking water limits. Radioactive and chemical contaminants in groundwater include, but are not limited to, tritium, iodine-129, Tc-99, uranium, Sr-90, nitrate, carbon tetrachloride, trichloroethene, and hexavalent chromium (Thompson 2007; Uziemblo 2007). There are 761 buried waste sites in the River Corridor project (i.e., sites located along the Columbia River) and 850 on the Central Plateau where the 200 East and 200 West Areas are located. There is also extensive contamination in the vadose zone at the site. Hanford groundwater does not directly serve as a source of potable water for municipal or private wells. However, Hanford groundwater does flow into the Columbia River where most downstream municipalities receive their drinking water. Mercury contamination is not a major issue at Hanford (Morse 2007; Uziemblo 2007).

DOE, EPA, and the Washington State Department of Ecology have developed a remediation plan for protecting the Columbia River Corridor. The Groundwater Remediation Project is largely responsible for ensuring the plan is implemented. The goals of the program are to prevent contaminated groundwater from migrating to the Columbia River, avoid groundwater contamination in the future, and remediate existing contamination (Jewell 2008). The primary groundwater contaminant plumes of concern are in the 100 and 300 Areas that adjoin the Columbia River. This is where former reactors were built, nuclear fuel development took place, and research laboratories were located. Contaminants of concern include chromium, strontium-90, and uranium with co-contaminants nitrate and trichloroethene (Thompson 2007). The primary groundwater contaminant plumes of concern for river protection in the 200 Area, located at the center of Hanford and generally associated with waste from plutonium reprocessing, are carbon tetrachloride, uranium, technetium-99, and iodine-129 (Thompson 2007). Approximately 16,000 m (or 16 km) of Columbia River shoreline receive contaminated groundwater migrating from beneath Hanford (Thompson 2007).

The Groundwater Remediation Project pays close attention to five practical actions:

- Remediate High-Risk Waste Sites—Clean up waste sites that pose the highest risk to groundwater;
- Shrink the Contaminated Area—Reduce the contaminated surface area, so as many areas as possible will no longer pose a threat to groundwater;
- Reduce Recharge—Reduce the transport of contaminants to groundwater from water released onto the soil;
- Remediate Groundwater—Complete remedial actions at pump-and-treat sites; and
- Monitor Groundwater—Determine the groundwater monitoring needs for long-term stewardship of the Central Plateau and evaluate new technologies that may be more effective (Jewell 2008).

Ongoing Columbia River Protection Projects include well-established groundwater control measures as well as technologies requiring applied or fundamental research. For example, the carbon tetrachloride plume is presently undergoing pump and treat, a method widely used for plume control. Vapor extraction of carbon tetrachloride began in 1992 followed by groundwater extraction 2 years later. An in situ reducing barrier was installed and began operating in 2002 to transform hexavalent chromium into the much less soluble Cr(III). A portion of the barrier is now losing reducing capacity and research into a method to mend the barrier is ongoing (Peterson 2007). Barrier systems offer plume containment, but do not actively remediate source zones. Because the latter is desirable, alternative methods to treat chromium and accelerate cleanup are also listed as needs.

Within the 200 Areas, the vadose zone varies in thickness from 50 m to 100 m and contaminants exist throughout the full thickness at various locations. Remediation methods for deep, unsaturated soils are not well known. Hanford, as well as at all of the other DOE sites visited by the committee, will rely on engineered caps and barriers to prevent water from carrying contaminants out of areas where solid wastes or contaminated materials are disposed onsite. A Hanford-designed prototype surface barrier, referred to as the Hanford barrier, is a 2.5-hectare multilayered, vegetated, capillary barrier composed mainly of stable natural materials and designed to isolate buried wastes for about 1,000 years. While not all near-surface disposals at Hanford will require the degree of protection offered by the Hanford Barrier, Ward (2007) stated that the results of tests and monitoring of the barrier's performance can be used to guide the design of more modest covers tailored to achieve the waste isolation needs of individual sites. To do so, it will be necessary to determine moisture flux through representative waste sites, including vegetated and graveled surfaces, account for seasonal variations in precipitation and heating, and from this information develop robust infiltration barriers for sites where contaminants will be temporarily or permanently left in place.

Thompson (2007) reported that a 2006 audit by the Government Accountability Office found fault with DOE's remediation efforts to prevent contaminants from reaching the Columbia River. The audit concluded that technology used in several remedies is not performing satisfactorily, and that there is a lack of new technologies to address contamination issues. Thompson also listed key contaminants that had raised congressional concerns in 2006. These were in two categories:

1. Contaminants currently entering the river—including hexavalent chromium in the 100 Area reactor sites, Sr-90 at N-reactor, uranium in the 300 Area, and tritium and I-129 from 200 East Area; and
2. Contaminants from the central plateau (200 Area) that may reach the river based on their half-lives, mobility, and inventory—including uranium, Tc-99, and carbon tetrachloride.

Stewart (2007) provided an overview of near-term geoscience challenges that PNNL addresses at Hanford. These include:

- The complex geohydrology at the groundwater–Columbia River interface and within the subsurface of the Central Plateau where most Hanford contaminants are in surface facilities, underground tanks, or already released into the ground;
- Hydrology and geophysics characterization and remediation of the deep vadose zone;
- Specific geochemistry issues, including Sr-90 around N-reactor, uranium in the 300 Area; and uranium, Tc-99, carbon tetrachloride, and plutonium in the Central Plateau; and
- Reactive transport modeling of contaminants in complex subsurface physical and geochemical settings.

To address these challenges, Stewart described several PNNL-led strategic initiatives aligned with site needs. Among the needs are cost-effective, in situ technologies to remediate chlorinated organics. Treatment of these species in deep vadose or tight (low-permeability) zones is particularly problematic. She also stated that mobile ions such as Tc-99, U, and Pu, which are prevalent at multiple sites, are difficult to treat in situ, especially in the deep vadose zone. Stewart noted that simple conceptualizations do not always adequately represent complex subsurface conditions; hence the need for a framework for translating science into conceptual and numerical models, as well as protocols for selecting and implementing numerical models to adequately address complexity of the geohydrologic environment. As long-term needs, she included cost-effective approaches to monitor residual contamination and to verify the performance of site remediation activities. She concluded that sampling and characterization technology, modeling, in situ technology, and long-term monitoring are common challenges at DOE sites (Stewart 2007).

From the perspective of the site cleanup contractor, Fluor Hanford, Peterson (2007) described the following challenges:

- Develop cost-effective in situ remediation of carbon tetrachloride and hexavalent chromium in the vadose zone,
- Develop cost-effective in situ remediation for radionuclides in the deep vadose zone,
- Develop numerical models that include chemical reactions by contaminants and their transport in groundwater and the vadose zone, and
- Develop improved, cost-effective methods for subsurface access to support characterization and remediation.

Roadmap Area: Decontamination and Decommissioning (D&D)

The D&D of the many inactive sites at Hanford has been a large part of the overall cleanup efforts that have been ongoing for the past 20 years at Hanford. The site is still facing a significant amount of D&D work, including 486 structures/facilities in the River Corridor project and over 950 structures/facilities on the Central Plateau (Romine 2007).

The River Corridor project includes Hanford's former production reactors, fuel fabrication, and laboratory facilities. For this project, most cleanup decisions have been made and cleanup has been initiated. For example, all reactor support structures are being removed, leaving only the reactor's core enclosure. As of 2007, four of Hanford's nine heavily reinforced concrete reactor buildings were "cocooned" to be left in place for 75 years to allow decay of most contaminants and future decisions as to their further D&D (NRC 2005). Hanford's B Reactor, the world's first full-scale operating reactor, is designated as a National Landmark and will be used as a Manhattan Project museum. It will not be cocooned like the other 8 reactors. The land in the river corridor will be available for other purposes such as conservation, tribal, recreational, and industrial use after cleanup (Romine 2007).

The 75-square-mile Central Plateau houses fuel reprocessing and waste management facilities, including the five very large "canyon" facilities used for reprocessing irradiated spent fuel from Hanford reactors. The Central Plateau is the last remaining major area where cleanup decisions are yet to be made. According to Romine (2007) the end-state assumptions are that:

1. The Central Plateau will remain under federal control indefinitely,
2. Institutional controls will remain in place for the foreseeable future, and
3. Legacy TRU-contaminated materials and soils will be left in place.

As an example of the D&D challenges posed by the canyons, the Purex canyon is approximately 1,000 feet long with walls up to 7 feet thick. Contamination and radiation levels preclude human entry into former fuel reprocessing cells (DOE 1997). There are also two tunnels near the Purex plant containing failed equipment, D&D, and other debris from Purex and the 300 Area located on 24 railroad cars (Gephart 2003). Contaminants include lead, mercury, cadmium, barium, plutonium, and miscellaneous fission products. The equipment contains solids and perhaps liquids. Perhaps 2 million curies of radioactivity exist in the tunnels. This will pose a challenge for D&D efforts. Records and access to the tunnels are not readily available (Hughes et al. 1994).

Romine (2007) stated that the baseline for canyon disposition is to seal the canyons in place rather than removing them. Given this plan, modeling efforts will be necessary to determine contaminants that can be disposed inside a canyon building without adversely impacting human health or the environment (Romine 2007). Other modeling work to support the baseline will be to investigate the long-term stability of the structures as well as grout added during the D&D. In addition Romine noted the needs for field screening methods to characterize hazardous and TRU contaminants in the canyons and for multipurpose cost-effective robotic vehicles to perform D&D tasks.

A major challenge for EM is to balance actual demolition work with long-term requirements for the larger, complex facilities. Engineering work for decommissioning a reactor could take 5 years to complete, while other facilities might only take 6 months. Most major D&D seems to have been pushed into the future. Reactors have been cocooned and reprocessing canyons stabilized with the hard and expensive part being in abeyance. For closure, the long-term performance of cement/grout is pivotal. These materials comprise the structure of facilities to be collapsed on themselves as well as possibly used to fill pipes, vessels, reprocessing cells, galleries, and other void spaces. There is a need to better understand the long-term performance of cement and the surface barriers that may eventually cover these facilities. Asbestos siding (transite) is also an issue on this site as on other DOE sites (Romine 2007).

Roadmap Area: Spent Fuel and Nuclear Materials

SNF

There are 2,100 metric tons of SNF, mostly N-reactor fuel, which is zirconium-clad uranium metal at Hanford. The plan is to send it to a deep geologic repository, such as DOE's planned Yucca Mountain repository. However, it is unclear that the fuel will meet the waste acceptance criteria for repository disposal and whether the repository will open. If not, other options for spent fuel management must be examined, including onsite storage and fuel reprocessing at Hanford or elsewhere (Gephart 2003). Each option will have unique science and technology needs.

K-Basin Sludge

Sludge that has been retrieved from K-basin arose from sloughing of uranium metal from N-reactor fuel stored for a prolonged time underwater in the basin, corroded fuel components and uranium, wind-borne soils that settled in the basin, and basin operations. The sludge is heterogeneous and contains a highly variable mix of chemicals, uranium metal, and other radionuclides. They are among the most dangerous materials on the Hanford site. The uranium metal is pyrophoric and reacts with water to produce hydrogen, and there is no efficient way to measure the amount of the uranium metal (Delegard et al. 2007).

The retrieved sludge has been containerized and is being stored in three locations in K-West area. Disposition plans for the sludge have changed five times since 1995. The plan at the time of the committee's visit, grouting and disposal in the WIPP, was in the conceptual design phase with additional needs to sample and characterize some of the sludge to determine if it can be accepted at WIPP (Delegard et al. 2007). The current plan is to repackage the sludge in new containers suitable for long-term storage away from the K-West Area while final treatment and disposal options are developed.

Cs-137 and Sr-90 Capsules

In 1968 Hanford's B-Plant began separating cesium and strontium from tank waste in order to reduce radioactive decay heat in the waste tanks and allow less-radioactive reprocessed tank liquids to be discharged into the soil. A new facility was added in 1974 to encapsulate these radionuclides, as CsCl and SrF_2, inside stainless steel cylinders. The cylinders were intended for use as radiation sources, for example, to sterilize food or medical equipment.

Of the 2,217 capsules originally produced, 1,936 are stored in a water pool at B-Plant. The capsules contain 130 megacuries of cesium-137 and strontium-90—about one-third of the total radioactivity on the Hanford site (NRC 2003). The radiation dose adjacent to each capsule is extremely high, more than 1 million rads per hour, which would give a person a lethal dose of radiation in less than 1 minute if standing within 3 feet of an unshielded capsule (Gephart 2003). A decision on how to dispose of the capsules has been deferred. One long-term plan proposes to package them for dry surface storage until they have mostly decayed, which would require some 300 years (about 10 half-lives). Another option could be to open the capsules and mix their contents with the liquid HLW stream to be vitrified in the WTP.

CAPABILITIES AND INFRASTRUCTURE AT PNNL

PNNL's total funding was about $760 million in FY 2007. Almost 60 percent of this total was provided by DOE in the areas of science, energy, environment, and national security (Davis 2007). Direct and indirect support from EM was about $91 million (Walton 2008). Presentations from Davis and Walton, as well as PNNL investigators cited in the previous section of this appendix, made it clear that PNNL is closely engaged in EM work relevant to Hanford and, to a lesser extent, the other EM cleanup sites. A recent effort led by PNNL scientists and several other national laboratories outlined many of the major scientific and technical challenges facing DOE across the cleanup sites along with opportunities through focused R&D to reduce the technical risk and uncertainty (Bredt et al. 2008).

Davis (2007) stated that a unique feature of PNNL is its fundamental strength in chemistry. Walton (2007) summarized PNNL's abilities to contribute to EM's cleanup program. Areas of technical capabilities include:

- Subsurface science,
- Chemical process engineering,
- Ecological science,
- Integrated assessment and risk analysis, and
- Environmental and human health and safety.

Walton (2008) presented a list of PNNL facilities that he judged would be relevant to meeting EM's future R&D needs. These are the following:

1. Environmental Molecular Sciences Laboratory (EMSL):
   - scientific investigations in biogeochemistry, waste, solution, and materials chemistry, and supercomputing for subsurface fate and transport simulations;
   - laboratory and bench-scale research; and
   - DOE SC user facility, state-of-the art scientific research and computing.

2. Life Sciences Laboratory:
   - subsurface flow-cell biogeochemical fate and transport research;
   - laboratory and bench-scale testing; and
   - radiological and nonradiological soils, solutions, and simulants.

3. Radiochemical Processing Facility:
   - soil and groundwater biogeochemical fate and transport research;
   - waste and process chemistry, physical properties, mixing, transport, separations, and immobilization;
   - laboratory and bench-scale testing with bench tops, hoods, and hot cells; and
   - category-II nuclear facility for highly radioactive spent fuels, tank waste, contaminated soils and solutions, as well as spiked simulants.

4. 336 High Bay:
   - process mixing, retrieval, gas retention and release (safety) testing;
   - laboratory and bench-scale testing with bench tops and hoods;
   - large pilot- and full-scale testing with supporting labs, bench tops, and hoods; and
   - non-radiological chemical and physical simulants.

5. Applied Process Engineering Laboratory (APEL):
   - mixing, chemical processing and filtration, waste forms, materials, physical and chemical properties, glass development and testing;
   - laboratory, bench, and small pilot-scale testing; and
   - non-radiological chemical and physical simulants.

6. Process Development Laboratory (PDLE, PDLW):
   - process mixing, slurry transport, chemical processing, and filtration;
   - large pilot- and full-scale testing (e.g. full-scale piping and pumps); and
   - nonradiological chemical and physical simulants.

Relevant to facility needs, Mauss stated that appropriate testing is the key to technology utilization. New technologies should be tested as part of an integrated system, not as individual components. They should be tested at applicable scales with appropriate wastes, and the effects on a new technology in downstream processes should be fully understood (Mauss 2007).

REFERENCES

Bagaasen, L. 2007. Example of integration of science and engineering on the Bulk Vitrification Project. Presented to the Committee on Development and Implementation of a Cleanup Technology Roadmap. Hanford, Richland, WA, November 1.

Bredt, P., A.R. Felmy, P.A. Gauglitz, D. Hobbs, S. Krahn, N. Machara, M. McIlwain, B.A. Moyer, A.P. Poloski, K. Subramanian, J.D. Vienna, and B. Wilmarth. 2008. Scientific Opportunities to Reduce Risk in Nuclear Process Science. PNNL-17699. Richland, WA: Pacific Northwest National Laboratory. Available at http://www.pnl.gov/main/publications/external/technical_reports/PNNL-17699.pdf.

Brouns, T. 2007. Challenges to cleanup: River Protection Project technology needs and technical challenges. Presented to the Committee on Development and Implementation of a Cleanup Technology Roadmap. Hanford, Richland, WA, October 31.

Colson, S.D., R.E. Gephart, V.L. Hunter, J, Janata, and L.G. Morgan. 1997. A Risk Based Focused Decision-Management Approach for Justifying Characterization of Hanford Tank Waste, Rev. 2. PNNL-11231. Richland, WA: Pacific Northwest National Laboratory.

Davis, M. 2007. Welcome: The National Academies Committee on Development and Implementation of a Cleanup Technology Roadmap. Presented to the Committee on Development and Implementation of a Cleanup Technology Roadmap. Hanford, Richland, WA, October 31.

Delegard, C.H., A.J. Schmidt, R.B. Baker, J.P. Sloughter, and W.W. Rutherford. 2007. Challenges to cleanup: K-basin sludge. Presented to the Committee on Development and Implementation of a Cleanup Technology Roadmap. Hanford, Richland, WA, October 31.

DOE (U.S. Department of Energy). 1997a. Integration Plan for Environmental Monitoring, Performance Agreement SM 7.2.1. U.S. Department of Energy. Fluor Daniel Hanford, Inc.

DOE. 2008. EM Science and Technology Roadmap: Reducing the Uncertainty in the EM Program. Washington, DC: DOE.

DOE-ORP/RL (Department of Energy Richland Operations Office and the Office of River Protection). 2007. Hanford Federal Facility Agreement and Consent Order: Tri-Party Agreement. Available at http://www.hanford.gov/?page=91&parent=0.

Gephart, R.E. 2003. Hanford: A conversation about nuclear waste cleanup. Presented to the Committee on Development and Implementation of a Cleanup Technology Roadmap, Hanford, Richland, WA, October 31.

Gephart, R.E. 2007. Overview of Hanford plutonium production and waste/nuclear material legacy. Presented to the Committee on Development and Implementation of a Cleanup Technology Roadmap. Hanford: Richland, WA, October 31.

Honeyman, J. 2007. Hanford HLW program—Challenges to cleanup. Presented to the Committee on Development and Implementation of a Cleanup Technology Roadmap, Hanford, Richland, WA, October 31.

Hughes, M.C., L.M. Douglas, and S.G. Marske. 1994. Accelerated Decontamination and Decommissioning at the Hanford Site. WHO-SA-2131-FP. Prepared for the U.S. Department of Energy. Richland, WA: Westinghouse Hanford Company. Available at http://www.osti.gov/bridge/servlets/purl/10118729-hIJzCh/native/10118729.pdf.

Jewell, G. 2008. Groundwater at the Hanford Site. U.S. Department of Energy. Available at http://www.hanford.gov/cp/gpp/. Accessed January 24, 2008.

Mauss, B. 2007. Hanford HLW program: Challenges to cleanup. Presented to the Committee on Development and Implementation of a Cleanup Technology Roadmap, Hanford, Richland, WA, October 31.

Michener, T. 2007. M-3: Mixing systems. Presented to the Committee on Development and Implementation of a Cleanup Technology Roadmap, Hanford, Richland, WA, October 31.

Morse, J. 2007. Hanford cleanup. Presented to the Committee on Development and Implementation of a Cleanup Technology Roadmap, Hanford, Richland, WA, October 31.

NRC (National Research Council). 2003. Improving the Scientific Basis for Managing DOE's Excess Nuclear Materials and Spent Nuclear Fuel. Washington, DC: The National Academies Press.

NRC. 2005. Improving the Characterization and Treatment of Radioactive Wastes for the Department of Energy's Accelerated Site Cleanup Program. Washington, DC: The National Academies Press.

NRC. 2006. Tank Waste Retrieval, Processing, and On-Site Disposal at Three Department of Energy Sites: Final Report. Washington, DC: The National Academies Press.

Peterson, S. 2007. Technology challenges for Hanford soil and groundwater remediation. Presented to the Committee on Development and Implementation of a Cleanup Technology Roadmap, Hanford, Richland, WA, October 31.

Romine, L.D. 2007. Deactivation and decommissioning. Presented to the Committee on Development and Implementation of a Cleanup Technology Roadmap, Hanford, Richland, WA, October 31.

Smith, R.I. 2007. Concerns about the current WTP system. Presented to the Committee on Development and Implementation of a Cleanup Technology Roadmap, Hanford, Richland, WA, October 24.

Stewart, T. (2007). Challenges to Cleanup: Soils and Groundwater Remediation. Presented to the Committee on Development and Implementation of a Cleanup Technology Roadmap, Hanford, Richland, WA, October 31.

Tamosaitis. 2007. Waste Treatment Plant Project: Technical accomplishments, needs, and challenges. Presented to the Committee on Development and Implementation of a Cleanup Technology Roadmap, Hanford, Richland, WA, October 31.

Thompson. 2007. Hanford Groundwater Overview—Challenges to cleanup: DOE-RL Groundwater Project. Presented to the Committee on Development and Implementation of a Cleanup Technology Roadmap, Hanford, Richland, WA, October 31.

Uziemblo, N. 2007. Washington State Department of Ecology Nuclear Waste Program: Hanford's Long-Term Cleanup Problems. Presented to the Committee on Development and Implementation of a Cleanup Technology Roadmap, Hanford, Richland, WA, October 31.

Walton, T. 2007. Enabling EM's cleanup mission with science and technology. Presented to the Committee on Development and Implementation of a Cleanup Technology Roadmap, Hanford, Richland, WA, October 31.

Walton, T. 2008. Critical PNNL Laboratories Supporting DOE EM. Presented to the Committee on Development and Implementation of a Cleanup Technology Roadmap, Washington, DC, April 28.

Ward, A. 2007. Prototype Hanford Surface Barrier. Presented to the Committee on Development and Implementation of a Cleanup Technology Roadmap, Hanford, Richland, WA, November 2.

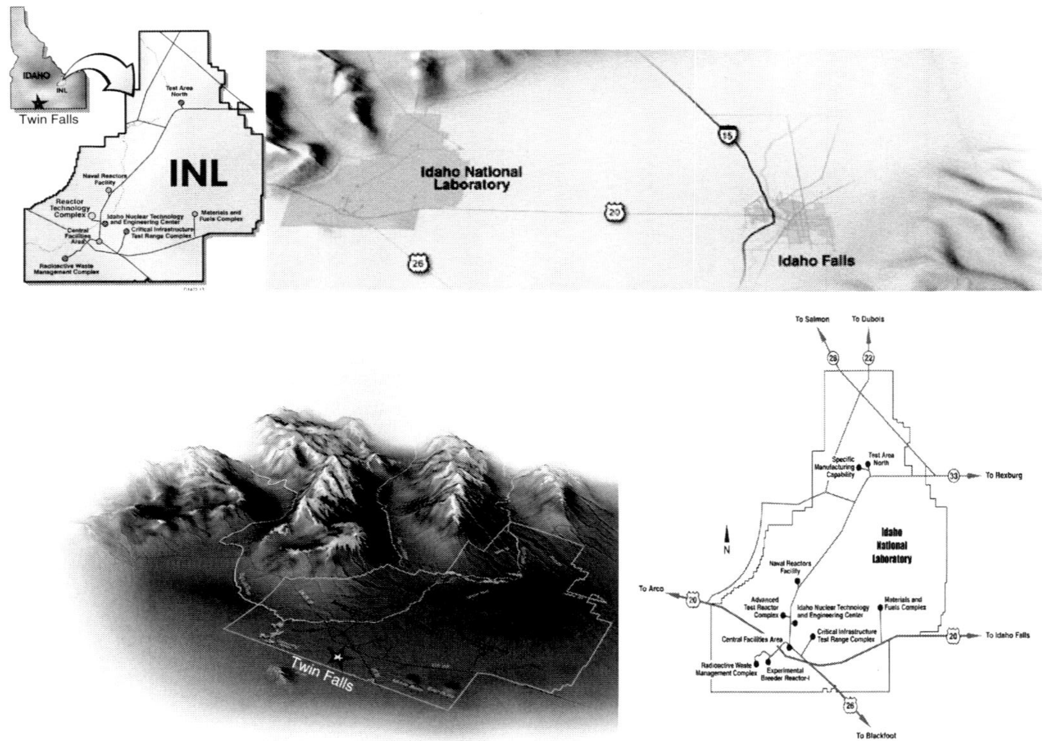

FIGURE E.1 The Idaho National Laboratory site encompasses 890 square miles in eastern Idaho. It was established in 1949 primarily for the development and testing of nuclear reactors, which continues to be a major mission for the site.
SOURCE: Department of Energy.

Appendix E

Idaho National Laboratory

INTRODUCTION

The National Research Council Committee on Development and Implementation of a Cleanup Technology Roadmap held its third meeting in Idaho Falls, Idaho from August 27 to 29, 2007. The purpose of the meeting was to obtain information relevant to the committee's Statement of Task (SOT) through presentations and tours by Department of Energy (DOE) staff and their contractors.[1]

This appendix provides a factual summary of the information related to the four items in the committee's SOT obtained during the meeting, the site visits, and documents provided to the committee. This appendix first describes the history and status of the DOE site in Idaho to provide perspective on the range of cleanup issues being managed by the DOE Office of Environmental Management (EM). The next sections summarize information presented to the committee, which guided the committee's deliberations in addressing its SOT as described in the main text. This appendix thus provides support for the findings and recommendations developed by the committee.

HISTORY

The Idaho National Laboratory (INL) site is the focus of EM's cleanup activities in Idaho. The site is located in the Idaho desert west of the city of

[1] The agenda for this meeting is shown in Appendix B.

Idaho Falls. The Idaho National Laboratory, which is associated with the site, also has offices and low-hazard laboratories in Idaho Falls.

The INL site is an 890-square-mile (569,135-acre) government reservation. It was established in 1949 as the National Reactor Testing Station and renamed the Idaho National Engineering Laboratory in 1977, the Idaho National Engineering and Environmental Laboratory in 1997, and INL in 2005. Fifty-two nuclear reactors were built on the site, including the U.S. Navy's first prototype nuclear propulsion plant, but most are no longer operated and many no longer exist. Nuclear fuel was reprocessed and wastes were managed through treatment, storage, disposal, or combinations thereof. During the 1990s, the laboratory's mission broadened into other areas, such as biotechnology, energy and materials research, and conservation and renewable energy. At the end of the Cold War, waste treatment and cleanup of previously contaminated sites became a priority. Today, INL is a science-based, applied engineering national laboratory dedicated to addressing national environmental, energy, nuclear technology, and national security needs, while cleanup continues under the separate Idaho Cleanup Project (ICP).

Begun in 2005, the ICP is a 7-year, $2.9 billion program funded through EM, which focuses equally on reducing risks to workers, the public, and the environment and on protecting the Snake River Plain Aquifer, the sole drinking water source for more than 300,000 residents of eastern Idaho. The cleanup contractor, CH2M-WG Idaho LLC (CWI)[2] has identified five major geographic areas at INL that are undergoing cleanup (CWI 2007):

- Idaho Nuclear Technology and Engineering Center (INTEC),
- Power Burst Facility (PBF),
- Reactor Technology Complex (RTC),
- Radioactive Waste Management Complex (RWMC), and
- Test Area North (TAN).

The following sections briefly describe the history and status of these geographic areas, followed by a short description of the structure and scope of the DOE-EM cleanup program at the INL (adapted from CWI 2007).

[2] CH2M·WG Idaho is a limited liability company formed by a partnership of *CH2M HILL* and the *Washington Division of URS Corporation* (formerly Washington Group International).

INTEC[3]

The Idaho Chemical Processing Plant was established in the 1950s to recover usable uranium in spent fuel from government reactors. Over the years, the facility recovered more than $1 billion worth of highly enriched uranium, which was returned to the government fuel cycle. A high-level liquid waste treatment process known as calcining was developed to reduce the volume of liquid waste generated during reprocessing and place it in a more stable granular solid form (CWI 2007). This waste, which currently contains about 44 million curies (MCi) of radioactivity, is to be immobilized in a form suitable for disposal in a high-level waste (HLW) repository and then shipped out of the state for disposal. Altogether some 900,000 gallons of waste, referred to as sodium-bearing waste, remain in 3 of 11 existing underground stainless steel storage tanks (Lockie 2007). These tanks are smaller than those at either the Savannah River Site (SRS) or Hanford and there is good access to most parts of the interior of the tanks. According to information given the committee, the Idaho tanks do not currently have any leaks.

The processing facility underwent modernization during the 1980s, including new structures to replace most major facilities. Nuclear waste reprocessing stopped in 1992, when DOE decided that reprocessing was no longer necessary. In 1998, the plant was renamed the INTEC.

Groundwater beneath INTEC has been impacted by historic operations of an injection well and disposal ponds, and by leaks in waste handling pipes and tanks over time. Treated wastes[4] from reprocessing spent nuclear fuel were injected into the aquifer from 1953 through 1984. Leaks in pipes and tanks and waste from other sources have resulted in contaminated groundwater perched above the aquifer. Contaminants found in the aquifer because of INTEC operations include tritium, iodine-129, strontium-90, technetium-99, sodium, chloride, and nitrate (IDEQ 2008).

Today, INTEC is focused on cleanup and protection of the Snake River Plain Aquifer from further contamination. The identified cleanup goals, on which some progress has already been made, are to:

- Transfer spent nuclear fuel from wet to dry storage and prepare for final disposition at an offsite repository;
- Treat liquid radioactive waste at the Integrated Waste Treatment Unit (IWTU);

[3] Adapted from https://idahocleanupproject.com/, the NRC (2005) report entitled Risk and Decisions about TRU and HLW, and http://www.deq.idaho.gov/inl_oversight/about/facilities/intec.cfm.

[4] These were non-HLW according to site information given to the committee.

- Characterize, repackage, and ship remote-handled transuranic (TRU) waste;
- Close liquid waste tanks;
- Remediate contaminated environmental soil sites; and
- Demolish and/or disposition excess facilities.

Progress made includes disposition of nuclear material items, transfer of approximately half of the spent nuclear fuel units from wet storage to dry storage in casks, and grouting seven 300,000-gallon HLW storage tanks. Approval has been received to begin construction of the IWTU, which is intended to treat the remaining high-level liquid wastes (CWI 2007).

PBF[5]

The PBF nuclear reactor was built in the 1970s and supported DOE and Nuclear Regulatory Commission studies of reactor fuel under both normal and off-normal operating conditions. The reactor was installed in a three-story, 19,000-square-foot facility. The PBF could subject fuel samples to large power surges in milliseconds, causing the fuel to fail in an isolated, contained system. Information obtained was used to help develop safe operating limits for commercial nuclear power plants. The PBF was shut down in 1998 and its nuclear fuel removed. The reactor vessel contains radioactive isotopes of cesium, strontium, and cobalt, and there are two highly contaminated cubicles in the first basement level of the facility.

The cleanup goals, which are in progress, include:

- Removal and disposition of lead, asbestos, and hazardous components;
- Disposition of the PBF reactor vessel; and
- Demolition of the containment facility and one nearby excess facility, completely eliminating the PBF footprint.

RTC[6]

The RTC, formerly Test Reactor Area, supports INL's nuclear energy research mission. Three major test reactors have operated at the RTC: the Materials Test Reactor (MTR, 1952-1970), the Engineering Test Reactor (ETR, 1957-1982), and the Advanced Test Reactor (ATR, 1967-present).

[5] Adapted from https://idahocleanupproject.com/.

[6] Adapted from https://idahocleanupproject.com/ and http://www.deq.idaho.gov/inl_oversight/about/facilities/tra.cfm.

The MTR was the second reactor to be operated at the INL site.[7] Data from the MTR influenced the choice of core structural materials and fuel elements for reactors subsequently designed in the United States. The ETR was larger and more flexible than the MTR and was used to evaluate fuels, coolant, and moderators under conditions similar to those in power reactors. The ATR is used to study the effects of radiation on materials and continues to support INL's energy research mission. This reactor also produces selected medical and industrial isotopes. Data from these reactors helped establish the technical bases for the design of subsequent reactors and the regulation of nuclear energy.

Past disposal of industrial, sanitary, and radioactive wastes to unlined ponds, and industrial wastes to injection wells, resulted in contaminated soil and groundwater perched above the Snake River Plain Aquifer and tritium, chromium, and sulfate contamination in the aquifer itself. Currently, low-level radioactive wastes are sent to a lined evaporation pond, and industrial and sanitary wastes are sent to infiltration ponds.

The ETR and MTR are scheduled for demolition as part of the site's 2012 cleanup. Although the nuclear fuel has been removed from both reactors, they still contain radioactive isotopes of cobalt, strontium, and cesium. The ETR contains an estimated 3,000 curies of cobalt-60. The ETR also contains tritium and low-concentration TRU contamination, and both reactors contain lead, graphite, and a total of more than 7,000 curies contained in irradiated beryllium. Cubicles in both facilities have contained more than 1 million pounds of lead and extensive asbestos-lined piping runs. A complication is that there are ATR utilities in the MTR basement, but care is being taken to ensure continuity of ATR operations.

The major cleanup goals, on which progress is being made, are to:

- Disposition the reactor vessels of the ETR and MTR, and
- Demolish ETR and MTR containment and support facilities.

Significant progress to date includes complete demolition of one building and partial dismantlement of others, plus removal of the ETR reactor vessel.

RWMC[8]

DOE has used the RWMC since the 1950s to manage, store, and dispose of radioactive waste generated in national defense and research

[7] Experimental Breeder Reactor No. 1, a Registered National Historic Landmark, was the first operating reactor at the INL site.

[8] Adapted from https://idahocleanupproject.com/ and http://www.deq.idaho.gov/inl_oversight/about/facilities/rwmc.cfm.

programs. Wastes originated from onsite operations as well as from other DOE operations, such as the Rocky Flats Plant in Colorado.

The Subsurface Disposal Area (SDA) is a 97-acre radioactive waste landfill that has been used for more than 50 years and is the major focus for remedial decisions and actions at the RWMC. Approximately 35 of the 97 acres contain waste from historical operations, including weapons production and reactor research. Most of the TRU waste was received from the Rocky Flats Plant prior to 1970 and buried at the SDA. The waste includes radioactive elements, organic solvents, acids, nitrates, and metals. Historical waste disposal practices have resulted in the release of radioactive and organic contaminants to the soil and groundwater below the SDA.

Targeted waste located at the SDA is identified, retrieved, and prepared for characterization and shipment to the Waste Isolation Pilot Plant (WIPP) in New Mexico under the Accelerated Retrieval Project. Enclosures are placed over sections of the pits where wastes are being accessed to isolate them from the environment.

The Transuranic Storage Area is primarily dedicated to managing contact- and remote-handled solid TRU waste prior to it being shipped offsite. The Advanced Mixed Waste Treatment Project (AMWTP), managed by Bechtel BWXT, Idaho, LLC is located here. The AMWTP is currently treating and shipping TRU waste to WIPP near Carlsbad, New Mexico, which is the nation's permanent deep-geologic repository for TRU waste.

The major cleanup goals, which are in progress, are to:

- Remove and dispose of targeted waste from specified portions of the SDA,
- Continue extracting organic vapors from the subsurface until remediation goals are met,
- Demolish excess facilities, and
- Ship TRU waste offsite.

TAN[9]

TAN was established in the 1950s in the northern portion of the INL site to support the government's Aircraft Nuclear Propulsion program. The goal was to build and fly a nuclear-powered airplane. Following cancellation of the nuclear propulsion program in 1961, other activities have been conducted at TAN.

The Loss of Fluid Test (LOFT) reactor, constructed between 1965 and 1975, was a scaled-down version of a commercial pressurized water

[9] Adapted from https://idahocleanupproject.com/ and http://www.deq.idaho.gov/inl_oversight/about/facilities/tan.cfm.

reactor. Loss-of-fluid experiments to simulate reactor fuel meltdowns were conducted under very controlled conditions within the LOFT dome, which provided containment. The resulting data were incorporated into the commercial reactor operating codes. Thirty-eight experiments were conducted within the facility, including several small loss-of-coolant experiments designed to simulate the type of accident that occurred at Three Mile Island (TMI). TAN also housed the TMI Unit 2 Core Offsite Examination Program that ended in 1990.

Volatile organic compounds (VOCs), radionuclides, and treated sanitary wastes were disposed of in an injection well at TAN from 1953 through the early 1980s. Groundwater beneath TAN is now contaminated with a range of VOCs, tritium, and strontium-90 (IDEQ 2008).

One of the main continuing missions at TAN is the manufacture of tank armor for the U.S. Army's battle tanks at the Specific Manufacturing Capability Project. The main cleanup goals, which are in progress, are to:

- Demolish 19 excess facilities, and
- Demolish two high-risk facilities (TAN-607 and LOFT reactor).

In addition to these goals, some soil areas contaminated with radionuclides and petroleum products also require remediation. The longest-term remediation activity will be the continued treatment of a contaminant plume in the aquifer below TAN. This action is to reduce VOC contamination in the aquifer to below maximum contamination levels using in situ bioremediation, natural attenuation, and pump-and-treat methods (IDEQ 2008).

TECHNOLOGY GAPS IDENTIFIED DURING THE MEETING WHICH DEFINE AREAS WHERE RESEARCH AND DEVELOPMENT (R&D) MAY BE NEEDED

The ICP includes the following (Leake 2007):

- TRU disposition, including sodium-bearing waste (SBW) treatment and operation of the IWTU;
- Calcine disposition;
- Low-level and mixed low-level waste disposition;
- Pperation of the advanced mixed waste treatment facility (AMWTF);
- Comprehensive Environmental Response, Compensation, and Liability Act (CERCLA) remediations and operation of the Idaho CERCLA disposal facility;

- Waste removal, decontamination and decommissioning (D&D), and remediation activities in the RWMC;
- D&D of large, contaminated structures;
- High-level tank and bin closures; and
- Disposition of spent nuclear fuels and special nuclear materials.

In addition to these, there are several facilities currently under the responsibility of the DOE Office of Nuclear Energy (NE) that may be transitioned to EM (Gill 2007b).

Cleanup is proceeding as agreed in a series of Records of Decision on, or ahead of, schedule. In general, the site personnel did not describe many areas requiring new technologies. Certain specific technological needs called out during the March workshop, the August site visit, or in the presentations and materials supplied during the site visit are listed in the categories below. For some categories, no technological needs were actually called out; in which case a summary of the cleanup approach that the committee was told was being adopted is provided.

WASTE RETRIEVAL AND TREATMENT

Calcine Retrieval and Treatment

Idaho's HLW, in the form of a calcine, is fundamentally different from the high-level tank waste stored at Hanford and SRS. At Idaho some 8-9 million gallons of acidic reprocessing waste were evaporated and reduced to 4,400 cubic meters of the calcine, which is stored onsite in 43 shielded bins within 6 binsets. The calcine consists of abrasive, hygroscopic, granular oxides ranging in size from about 0.2 to 0.6 milimeters plus about 15 percent fine particles. There is significant heterogeneity in composition of the calcine due to layering in any given bin. INL has demonstrated technical approaches that can remove the dry calcine from the onsite storage bins (Hagers 2007).

According to Hagers (2007) the key issue is whether the calcined wastes in their present form, suitably packaged for transportation and geologic storage, will meet regulatory requirements for disposal at a high-level radioactive waste repository. If this is not the case, INL has three alternative paths (1) direct vitrification, (2) hot isostatic pressing (HIP), or (3) dissolution and steam reforming. DOE intends to issue a Record of Decision by the end of 2009 to identify a method to treat the calcine (if necessary). December 31, 2035 is the target date for having all calcine packaged and ready to ship out of state. INL has established a Calcine Disposition Project, which is part of the ICP, in order to meet these milestones.

Hagers' presentation emphasized the HIP process, noting its advantages in reducing final waste volumes, as well as its flexibility as a "finish" for several varieties of waste. DOE-Idaho currently has a HIP unit installed in a hot cell at INL, which will be used to perform trials on glass-ceramics and multiphase ceramics with nonradioactive simulants. Further requirements to demonstrate the HIP process for calcine are related to process throughput: in-line heater technology, can filling rate, and cycle time (Hagers 2007).

SBW Treatment

Idaho has about 900,000 gallons of tank waste that have accumulated from a variety of sources since nuclear fuel reprocessing ended in at INTEC in 1991 (Lockie 2007; Olson 2007). Referred to as SBW the highly acidic waste is stored in three tanks. Because of its acidity, the tank contents are relatively homogeneous (essentially all the SBW remains in solution with little precipitation) and differ fundamentally from the alkaline tank wastes at Hanford and Savannah River.

Within the ICP's SBW Project, over 100 technical options for dealing with the waste were considered. CH2M-WG Idaho proposed steam reforming as the preferred treatment technology and received the SBW contract in 2005 (Olson 2007).

The proposed steam reforming process takes place in a fluidized bed chemical reactor that operates between 625 and 740°C. Waste solution is atomized into the reactor where water is evaporated, and organics and nitrates or nitrites are converted to carbon dioxide, water, and nitrogen gas. Alkali metals and other inorganic constituents are incorporated into a granular product (Olson 2007).

Olson (2007) noted that fluidized bed steam reforming (FBSR) had undergone testing and demonstration by a number of commercial vendors and DOE laboratories. Bench-scale tests using a 6-inch-diameter fluidized bed were conducted at the Science Applications International Corporation's Science and Technology Applications Research center from 2003 through 2004. Hazen Research Inc., Colorado, conducted engineering-scale tests and demonstrations from 2005 through 2006. The proposed process is based on a Thor Treatment Technologies flow sheet.

FBSR was demonstrated for application to Hanford's low-activity tank waste at Pacific Northwest National Laboratory and Savannah River National Laboratory under a 2001 contract with Bechtel National, Inc. FBSR was also demonstrated for SRS Tank 48H waste solution in 2003. The engineering-scale tests by Hazen Research, Inc., have included both SBW and SRS Tank 48 waste simulants.

According to Olson (2007), FBSR of SBW appears to meet all processing criteria and results in a product that should meet geologic storage requirements. All SBW will be processed using this methodology and will be removed from the site by 2012. Olson (2007) concluded that FBSR is a robust treatment technology with broad applicability to waste streams throughout the DOE complex.

Tank Closure

Tank closure and grouting is ongoing at Idaho, with final closure of the tank farm scheduled for 2012 (Lockie 2007). With the exception of its SBW, INL converted its acidic reprocessing waste to calcine, thus avoiding the long-term tank storage of multiphase alkaline waste, which is the practice at Hanford and SRS. The amount of reprocessing waste that arose at INL was about 10 percent of that at Hanford and SRS (NRC 2005).

Idaho's tank farm system consisted of 11 underground, 300,000-gallon stainless steel tanks. Eight of these were constructed with cooling coils to remove decay heat from highly radioactive wastes; three have no cooling coils. In addition, four smaller, 30,000-gallon stainless steel tanks were used for storage, but taken out of service in the early 1980s. Among all of the tanks, only four 300,000-gallon tanks are still in service storing the SBW; the others having been emptied (Lockie 2007).

Lockie (2007) presented photographs and data indicating that tank cleaning has been effective. The residual radioactivity in the four smaller tanks (WM-103 through -106) amounts to about 144 curies. In the larger tanks (WM-180 through -186) the residual radioactivity averages about 1,000 curies in each. Cs-137 and its short-lived daughter isotope Ba-137 account for about 95 percent of the total radioactivity.

From November 2006 through March 2007 all four of the cleaned 30,000-gallon tanks were closed by being filled with grout. From April through July 2007, INL completed engineered placements of grout in seven cleaned 300,000-gallon tanks. The purpose of engineered placements is to push residual tank waste into configurations on the tank floor from which more waste can be recovered, and to ensure any remaining waste is well encapsulated in the grout. From July through August 2007, the time of the committee's visit, additional grout pours were taking place to completely fill the seven tanks. INL expects that the SBW Treatment Project will empty the remaining four tanks by 2010. Piping associated with the tanks will also be grouted. Lockie (2007) concluded that no technology gaps have been identified that would prevent completion of tank closure of the tank farm facility by 2012.

Remediation in the SDA of the RWMC

The SDA, which is located within the RWMC, encompasses some 97 acres in total and contains 35 acres with buried waste, including TRU waste (pre-1970) and low-level waste deposited from 1970 to the present (Arenaz 2007). There are nearly 425,000 containers of mixed waste, 230,000 of which are from the former Rocky Flats plant (deposited pre-1970).

DOE's remedial investigation/baseline risk assessment determined that the baseline risk (without remediation) is not acceptable. According to this assessment 12 radionuclides and 6 nonradionuclides pose unacceptable risk to human health and the environment based on a 1,000-year simulation period (Arenaz 2007). Following this assessment a feasibility study outlined preliminary remediation goals and evaluated a range of remedial alternatives. These alternatives range from taking no action to complete removal of all source terms at the SDA. The alternatives are as follow:

- No action, but continued environmental monitoring using the existing network;
- Emplacing a modified Resource Conservation and Recovery Act (RCRA) Type C surface barrier;
- Emplacing an evapotranspiration surface barrier;
- In situ grouting of specific disposals that contain mobile Tc-99 and I-129;
- Partial retrieval of targeted wastes from four acres of the disposal pit area;
- Partial retrieval from two acres of pit area with grout slurry walls installed around the perimeter of the pit area; and
- Full retrieval of targeted waste.

According to Arenaz (2007) the next steps are to develop a proposed plan, which might involve some combination of the above alternatives, consider public comments on the proposed plan, and draft a Record of Decision.

During the site tour, site personnel also noted that improved technologies for excavator equipment as well as personal protective equipment (PPE) would help make the waste excavation activities at RWMC safer and faster.

Soil and Groundwater Cleanup

Site research to date indicates that the aquifer under the Idaho reservation comes to the surface in the Twin Falls, Idaho area. Within the boundaries of the Idaho reservation, the aquifer is contaminated with radionuclides, hazardous chemicals (including organics [light nonaqueous phase liquids and dense nonaqueous phase liquids]), and human waste. The committee heard two presentations that dealt with soil and groundwater cleanup. One dealt with a large contaminant plume in TAN (Lee 2007) and the other dealt with contaminants in the vadose zone at the RWMC (Arenaz 2007).

Industrial wastewater was injected directly into the aquifer beneath TAN from 1953 to 1972. This has resulted in contamination in the aquifer, primarily trichloroethylene (TCE), between 200 to 400 feet deep, and a two-mile-long TCE plume. The aquifer is composed of fractured, basalt lava flows with interlayered sedimentary units deposited during periods of volcanic quiescence. INL has implemented a three-component remediation strategy, which includes:

- In situ bioremediation of the "hot spot" where the TCE concentration is greater than 20,000 micrograms per liter,
- Pump and treat in the medial zone where the TCE concentration is from 1,000 to 20,000 micrograms per liter, and
- Monitored natural attenuation in the distal zone where the TCE concentration is below 1,000 micrograms per liter.

Lee (2007) reported research to improve the bioremediation strategy of the hot spot. Accomplishments thus far include increased dissolution of the source material to make it more available for biodegradation, increased mass of the microbe population capable of degrading the TCE around the source area, and increasing the biological activity surrounding the residual source area.

The research also includes evaluating alternative remediation technologies in the medial zone. These evaluations include pump-and-treat, biological attenuation, and in situ biological degradation (Lee 2007).

Groundwater monitoring in the vicinity of the RWMC includes 23 monitoring wells drilled into the aquifer. In 1987 a variety of chlorinated VOCs were found in the groundwater. The VOCs included carbon tetrachloride, TCE, chloroform, trichloroethane, and tetrachloroethylene (perchloroethylene). In 1994 a CERCLA Record of Decision identified vapor vacuum extraction, with treatment to destroy the extracted VOCs, as the preferred method of remediation. Arenaz (2007) reported that the system had destroyed a total of over 100,000 pounds of VOCs through mid-2007, and that the system will continue operating.

D&D

TAN was built between 1954 and 1961 to support the Aircraft Nuclear Propulsion Program. It was subsequently converted to support a variety of DOE-ID research projects. TAN encompassed several facilities including

LOFT. The TAN Hot Shop contained the world's largest remote-handling facility for radioactive materials. TMI core debris was shipped to the TAN Hot Shop where examination revealed signs of melting (Shaw 2007).

Shaw (2007) described demolition of the TAN-650 containment dome of the LOFT. By 2012 the entire facility will be removed to ground level, with a cap placed over the remaining below-surface structure. Shaw (2007) also described demolition of the TAN-607 Hot Shop. Demolition was accomplished by cutting arches in its thick concrete shielding walls and then collapsing the remaining structure with explosive charges. This demolition was done without means to contain the resulting dust or debris.

The Hot Shop was the last major facility to be demolished at TAN. Since April 2005, all 44 excess facilities have also been demolished. In July 2008, CWI completed the TAN project, 4 years ahead of schedule and significantly under budget.

A large amount of mercury was reportedly used and lost during reactor testing at TAN. Site investigations have found mercury at TAN along rail tracks and in sumps. It is evidently uncertain if there is mercury under floors of facilities like the TAN-607 Hot Shop or similar areas. Some site personnel suspect that mercury contamination might be a significant problem.

Gill (2007b) described a list of NE-owned facilities that are proposed for transfer to EM during the period from 2009 through 2012. The most challenging of these appears to be the experimental breeder reactor (EBR-II) and its associated facilities. EBR-II was constructed to demonstrate a complete breeder-reactor power plant with onsite fuel reprocessing. It operated from 1964 to 1969. The reactor is shut down and defueled. Sodium-bearing coolant remains in the reactor coolant loops and other components.[10]

At INTEC, 112 excess facilities and 4 more high-risk facilities (CPP-648, CPP-601, CPP-603A, and CPP-640) are slated for deactivation, decontamination, and demolition (Leake 2007). By the time these latter facilities undergo D&D they will likely have deteriorated, which could make entry by workers hazardous.

Also at INTEC, extensive cleanup of the Flourinel Dissolution Process hot cell located in building CPP-666 is required to support the Remote-Handled Waste Disposition Project (Jines 2007).

D&D Worker Safety

Hain (2007) described two needs for improving worker safety during D&D and other work in areas where there is radiation or contamination:

---

[10] At the time this report was in review, the committee was informed that a panel of experts from within the United States and Europe was to convene in February 2009 to evaluate alternatives for removal and remediation of the residual sodium inventory while minimizing secondary waste generation (personal communication from Jay Roach, INL).

• Technology to support work in high radiation areas, including enhanced radiological monitoring, especially against high background levels: Specifically, there is a need for improved rejection of the signal from naturally occurring radon decay products in air monitors used for alpha-particle-emitting airborne contaminants; and

• Personal protection equipment (PPE) designed for high temperatures and longer exposures to the radiation environment: Currently available PPE is often too heavy and bulky, resulting in limitation of motion, extra exertion, and overheating. PPE with externally supplied air can have problems with the supply hose. Workers who can operate excavation equipment in the RWMC for only short periods of time due to heat stress were one example of this need, which was pointed out during the committee's site tour.

Site personnel also noted the need for improved removal, handling, and disposal methods for asbestos in and on buildings, especially transite, an asbestos-containing material for wall panels, which is to be removed from higher elevations on the sides of buildings.

SPENT NUCLEAR FUEL (SNF) AND SPECIAL NUCLEAR MATERIALS (SNM)

As a part of the Idaho Cleanup Project (ICP), Idaho's Materials Disposition Project (MDP) includes: (1) dispositioning by September 30, 2009, all SNM owned by the ICP and (2) managing all SNF and SNF facilities at INTEC and at Ft. St. Vrain, Colorado (Hain 2007). In 2005, the ownership of all SNM at Idaho was divided between DOE-NE and the ICP, which is under the responsibility of EM. Most ICP-owned SNM is unirradiated fuel (i.e., not SNF), and it is being rapidly dispositioned. Hain (2007) stated that the MDP has no technology needs related to SNM.

The MDP also manages legacy SNF from DOE, Department of Defense, foreign and domestic research reactors, and commercial reactors. There are some 220 fuel types including aluminum- stainless steel- and zirconium-clad fuels. Much of this fuel was stored underwater in the CPP-666 basin at the time of the committee's visit. Hain (2007) presented both near- and long-term research needs she judged important for safely managing SNF and related facilities at INTEC and Ft. St. Vrain, and to achieve Idaho Settlement Agreement SNF and RCRA Site Treatment Plan requirements, as follow:

Immediate need:

- portable method of confirming uranium content of SNF received at INTEC. Since the best time and place to confirm content is during preload inspection, such a method needs to be portable, capable of operations in various small spaces, and applicable to all fuel types.

Mid-term needs:

- technology to effectively dry and to confirm the dryness of SNF in a basket or container, and
- technology to drill into containers and provide internal inspection in high-radiation fields.

Post-2012 needs:

- efficient, nondestructive characterization methods for SNF to support repository acceptance,
- support for packaging/storage facility design, and
- enhanced high-field radiological monitoring.

Several SNF research needs are also relevant to managing other radioactive material and wastes. These include:

1. Continued improvements in radiation control during inspection, transport, and handling;
2. Continued improvements in crane design and other remote operated/robotic equipment:
   - For SNF management, improvements in remote manipulator and crane design, including "Design for Reliability" and "Design for Maintainability" enhancements are needed. The range of fuel types and the fact that some older fuel designs did not include "grapple" positions makes improvements in manipulations an important factor in reducing handling costs. Recent Navy fuel operations involving welding of fuel canisters for disposal in a geologic repository have identified needs for similar improvements in the remote systems used for fuel packaging.
   - For D&D and waste retrieval, the primary failure modes related to poor maintainability and equipment failure include clogging by the dust/effluvium inherent in the excavation environment and vibration from both continuous and periodic (impact) operation.
   - Site operators mentioned material failure (cracking) and electronic problems due to vibration that seriously impact the lifetime of equipment. It is also very difficult to perform maintenance or failure analysis of components (to determine root causes and corrective options for redesign) on equipment used in a high-radiation environment, leading to more slowdowns. This was cited specifically in WMF 1612 but, most likely, is a sitewide issue. Technology development in the area of vibration isolation materials and devices would also be of benefit.
3. Continued improvements in transport tracking:

- Difficulties in tracking of waste containers can be a very serious problem, resulting in delay of operations; site personnel described a need for improved capability for reliable reading (visibility) of bar codes; and
- Institution of paperless records management and acceptance of that approach in the Yucca Mountain quality assurance requirements would reduce the currently high records-storage costs and improve records retrieval in support of waste disposal at WIPP and Yucca Mountain.

Work relevant to the need for characterizing SNF in a high-radiation field was reported by McIlwain (2007). According to his work, $LaBr_3$:Ce is the optimium scintillator material for such application, and he has constructed a detector with optimized design for background suppression. He noted that such a detector may have applications for both WIPP and the planned Yucca Mountain repository.

Challenging Materials

In 2004, beryllium (Be) was declared as a "waste with no path to disposition." This is a result of the recognition that the beryllium reflector blocks removed from the ATR were contaminated with cobalt-60, carbon-14, tritium, and TRU elements (as a result of initial uranium impurities in the Be). According to Gill (2007a) the research needs related to Be fall into two categories: preirradiation for the construction of replacement reflector blocks for the ATR without the impurities that generate Co-60 and TRU by-products, and postirradiation for how to handle the older Be reflector blocks that contain Co-60, C-14, and TRU.

Preirradiation needs include:

- A source or purification process for Be that will result in low uranium content (to limit irradiation-formed TRUs) and low nitrogen content (to limit irradiation-formed C-14) of incoming Be stock, and
- Improved strength and stiffness to prevent or slow down swelling or cracking, thereby extending the life of Be components in a reactor.

Postirradiation needs are:

- There is no ongoing research on the processes required for separating the Be from the radioactive contaminants in used Be reflectors. The chlorine dissolution process followed by separation needs research. The Be recovered from this process might be used for new reflectors, thus providing higher-purity Be of low contaminant content; and

- Waste stabilization using vitrification needs to be examined for Be.

During the presentation, Gill noted that Be buried in the RWMC is a source of the tritium plume from that facility.

OPPORTUNITIES FOR LEVERAGING R&D BEING SPONSORED BY ORGANIZATIONS OTHER THAN EM AT THE INL

INL underwent a restructuring beginning in 2003, and the national laboratory was established with its current mission and organization in 2005. At that time R&D for cleanup was separated from site cleanup contracts. However, the laboratory is responsible for sitewide stewardship following cleanup. At the time of the committee's visit, INL expected about $50 million in funding for energy and environmental R&D in FY 2007 within a total budget of about $707 million. The laboratory expected about $16 million from EM (Connolly 2007). In his presentation, Connolly noted that R&D at national laboratories now looks quite expensive to cleanup contractors, and that there is little incentive for contractors to use the national labs.

Connolly (2007) reviewed INL's major program areas, with emphasis on where EM might leverage research with these programs, as follow:

1. Nuclear science and technology:
   - R&D for Generation IV (GEN IV) reactors,
   - nuclear fuel cycle development, and
   - modeling and simulation.
2. National and homeland security:
   - nuclear fuel cycle,
   - active interrogation systems (e.g., for detecting SNM), and
   - communication systems, wireless technology, and sensors.
3. Energy and environment:
   - environmental science,
   - biotechnology,
   - artificial intelligence for robotics, and
   - actinide chemistry.

Currently funded DOE Laboratory Directed Research and Development (LDRD) activities and Office of Biological and Environmental Research (OBER) activities, as well as funding that can be leveraged through contractor-directed research activities, has potential for improving subsurface characterization and remediation technologies. LDRD and OBER R&D activities aimed at developing tools that can detect *in situ* biological activity (e.g., quantitative polymerase chain reaction, fluorescent hybridization, enzyme activity probes) can assist in supporting decisions to allow the monitored natural attenuation, or accelerated attenuation, of organic plumes (e.g., the TCE plume at TAN) and also, potentially metals contamination on the site. Development of substances that can stimulate the growth of remediating bacteria can also advance bioremediation activities on the site. Lee (2007) described R&D, partly funded through a site contractor, that aided in a decision to replace lactate with whey protein in the bioremediation project at the TAN. According to Lee's presentation this decision appears to have helped improve the bioremediation activities for the TCE plume at TAN.

Hagers (2007) described a collaboration between the INL and the Australian Nuclear Science and Technology Organization to develop a new HIP immobilization technology for calcine disposition.

EXPERTISE AND INFRASTRUCTURE AT INL THAT MAY BE RELEVANT TO ADDRESSING THE R&D NEEDS OF THE EM CLEANUP PROGRAM

Connolly (2007) reviewed core capabilities at INL that could be useful to address future EM R&D needs. Many of these capabilities are currently associated with other INL programs. These programs could provide cooperative or "leveraged" R&D opportunities for EM as noted above. Laboratory capabilities are as follow:

1. Waste processing:
   - basic science and technology including:
     - computational chemistry,
     - coordination and separations chemistry,
     - molten salt chemistry,
     - radiochemistry and trace element analysis, and
     - thermodynamics of aqueous, nonaqueous, and ionic liquids.
   - solvent extraction-based separations,
   - thermal processing and immobilization with cold-crucible induction melter technology,
   - immobilization using HIP capabilities,
   - fluidized bed calcination and steam reforming, and
   - advanced fuel cycle and Global Nuclear Energy Partnership (GNEP) programs that are synergistic with EM waste processing.

2. Groundwater and soil remediation:[11]
   - geochemistry:
     – chemistry of surfaces and adsportion on mineral surfaces,
     – molecular-level interpretation of geochemical reactions, and analytical tools for their investigation, and
     – geochemistry of radionuclides.
   - modeling:
     – multiphase fluid flow,
     – solute modeling, and
     – performance assessment modeling.
   - characterization:
     – soil and rock physical properties,
     – hydrologic properties, and
     – autonomous monitoring.
3. D&D:
   - EM and NE have established a joint INL program to address the problem of removing metallic sodium from EBR-II.
4. SNF:
   - remote canister welding and nondestructive examination, and
   - remote handling of SNF.

The infrastructure that the committee observed during its site visit, and which might be relevant to addressing the R&D needs at Idaho and the other DOE sites, includes the following:

- The hot cell capabilities and the remote handling capabilities,
- The test reactor to provide high neutron fluxes to generate samples for test work,
- The CPP-666 fuel storage basin facility,
- The waste compaction ("super compactor") facility,
- The test area for grout fill applications,
- The FBSR project developed for treatment of SBW,
- The radiological calibration laboratory and the radiation detection laboratory,
- Facilities within the Advanced Separations and Radiochemistry department,
- The cold crucible research facility,
- The geo-centrifuge facility, and
- The AMWTP.

[11] Connolly (2007) noted that spatial dimensions of these capabilities range from the pore scale, through laboratory scale, and field scale.

REFERENCES

Arenaz, M. 2007. Waste Area Group 7—Operable Unit 7-13/14. Presented to the Committee on Development and Implementation of a Cleanup Technology Roadmap, Idaho National Laboratory, Idaho Falls, August 29.

CWI (CH2M-WG Idaho, LLC). 2007. Idaho Cleanup Project: Progress Report 2007. Available at https://idahocleanupproject.com/.

Connolly, M. 2007. INL overview Presented to the Committee on Development and Implementation of a Cleanup Technology Roadmap, Idaho National Laboratory, Idaho Falls, August 29.

Gill, R. 2007a. Beryllium stabilization and disposition. Presented to the Committee on Development and Implementation of a Cleanup Technology Roadmap, Idaho National Laboratory, Idaho Falls, August 29.

Gill, R. 2007b. Contaminated and noncontaminated NE-owned facilities proposed for transfer to EM during the period of 2009 through 2012. Presented to the Committee on Development and Implementation of a Cleanup Technology Roadmap, Idaho National Laboratory, Idaho Falls, August 29.

Hain, K. 2007. Idaho Cleanup Project—SNF management and disposition; special nuclear materials disposition. Presented to the Committee on Development and Implementation of a Cleanup Technology Roadmap, Idaho National Laboratory, Idaho Falls, August 29.

Hagers, J. 2007. High Level Waste (Calcine) Disposition Project. Presented to the Committee on Development and Implementation of a Cleanup Technology Roadmap, Idaho National Laboratory, Idaho Falls, August 29.

IDEQ (Idaho Department of Environmental Quality). 2008a. About INL Facilities: Idaho Nuclear Technology and Engineering Center (INTEC). Available at http://www.deq.idaho.gov/inl_oversight/about/facilities/intec.cfm.

IDEQ. 2008b. About INL Facilities: Test Area North (TAN). Available at http://www.deq.idaho.gov/inl_oversight/about/facilities/tan.cfm.

Jines, A.T. 2007. Introduction to the Remote-Handled Waste Disposition Project (RWDP). Presented to the Committee on Development and Implementation of a Cleanup Technology Roadmap, Idaho National Laboratory, Idaho Falls, August 29.

Leake, B. 2007. Idaho Cleanup Project. Presented to the Committee on Development and Implementation of a Cleanup Technology Roadmap, Idaho National Laboratory, Idaho Falls, August 29.

Lee, M.H. 2007. Development and evaluation of innovative remediation tools: Improvements to long-term performance and cost at the Test Area North groundwater plume. Presented to the Committee on Development and Implementation of a Cleanup Technology Roadmap, Idaho National Laboratory, Idaho Falls, August 29.

Lockie, K.A. 2007. Tank closure progress at the Department of Energy's Idaho National Engineering Laboratory Tank Farm Facility. Presented to the Committee on Development and Implementation of a Cleanup Technology Roadmap, Idaho National Laboratory, Idaho Falls, August 29.

McIlwain, M.E. 2007. Advanced radiation detector development for assay of transuranic elements in highly radioactive waste. Presented to the Committee on Development and Implementation of a Cleanup Technology Roadmap, Idaho National Laboratory, Idaho Falls, August 29.

NRC (National Research Council). 2005. Risks and Decisions about Disposition of Transuranic and High-Level RadioactiveWaste. Washington, DC: The National Academies Press.

Olson, A. 2007. IWTU Sodium-Bearing Waste (SBW) Treatment Project. Presented to the Committee on Development and Implementation of a Cleanup Technology Roadmap, Idaho National Laboratory, Idaho Falls, August 29.

Shaw, R.M. 2007. Test Area North decontamination and decommissioning (D&D). Presented to the Committee on Development and Implementation of a Cleanup Technology Roadmap, Idaho National Laboratory, Idaho Falls, August 29.

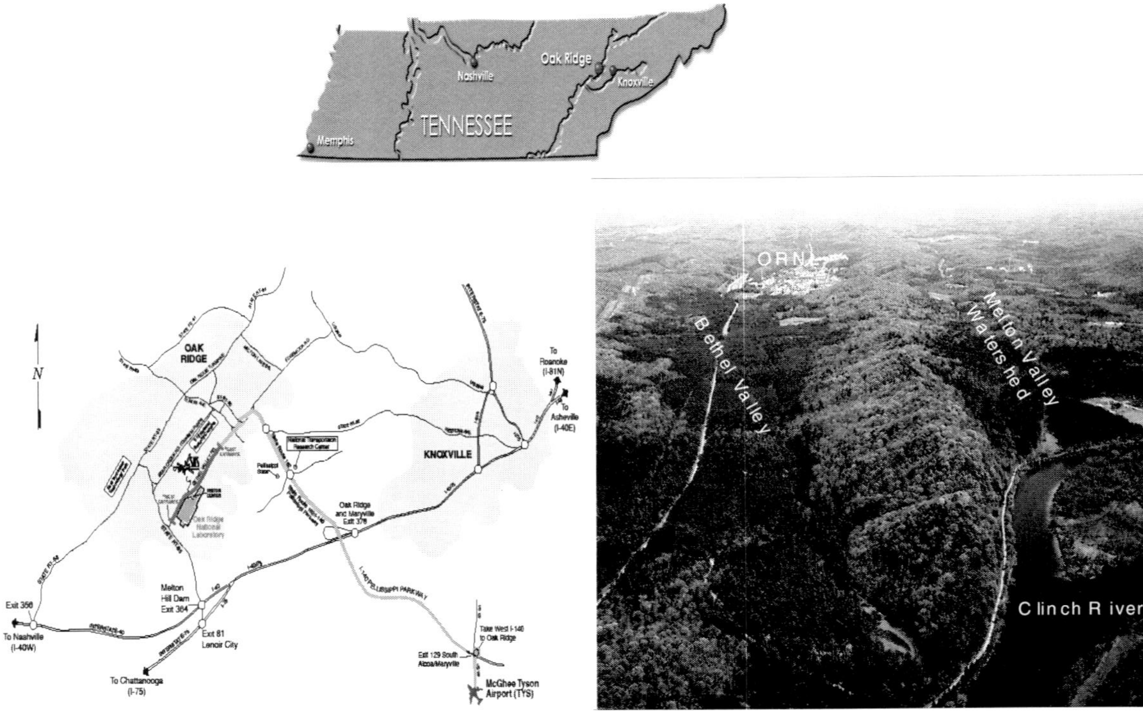

FIGURE F.1 The Oak Ridge Reservation includes over 30,000 acres within the city limits of Oak Ridge, Tennessee. Massive facilities for enriching uranium were built in 1943. Site operations continue today in the East Tennessee Technology Park, the Y-12 National Security Complex, and the Oak Ridge National laboratory.
SOURCE: Department of Energy.

Appendix F

Oak Ridge Reservation

INTRODUCTION

The National Research Council Committee on Development and Implementation of a Cleanup Technology Roadmap held its second meeting in Oak Ridge, Tennessee on June 13-15, 2007. The purpose of the meeting was to obtain information relevant to the committee's Statement of Task (SOT) through presentations and tours by Department of Energy (DOE) staff and their contractors.[1]

This appendix provides a factual summary of the information related to the four items in the committee's SOT obtained during the meeting, the site visits, and documents provided to the committee. This appendix first describes the history and status of the DOE site at Oak Ridge to provide perspective on the range of cleanup issues being managed by the DOE Office of Environmental Management (EM). The next sections summarize information presented to the committee, which guided the committee's deliberations in addressing its SOT as described in the main text. This appendix thus provides support for the findings and recommendations developed by the committee.

HISTORY

The DOE's activities in Oak Ridge are conducted on the Oak Ridge Reservation (ORR) which encompasses 33,750 acres within the city limits of Oak Ridge. The ORR contains three major sites (Figure F.1):

[1] The agenda for this meeting is shown in Appendix B.

- East Tennessee Technology Park (ETTP),
- Y-12 National Security Complex, and
- Oak Ridge National Laboratory (ORNL).

The following sections briefly describe the history and status of these sites. This is followed by a short description of the structure and scope of the DOE EM cleanup program at the ORR.

ETTP

The ETTP, which was formerly called the Oak Ridge Gaseous Diffusion Plant, was constructed in 1943 to produce enriched uranium hexafluoride for defense purposes and later for nuclear power reactors. The Gaseous Diffusion Plant produced highly enriched uranium (HEU) until 1964. It then switched to producing low-enriched uranium for use as fuel in commercial power reactors. In 1985 reduced demand resulted in the closing of the gaseous diffusion cascades. There remained an estimated 1.5 metric tons of HEU at the ETTP. Almost all of the HEU is contained in the K-25 Building at ETTP in the form of deposits on the internal surfaces of the shutdown processing equipment. All of the depleted uranium hexafluoride tails stored in cylinders have been moved to other sites.

ETTP now serves as the center of operations for DOE's Oak Ridge Environmental Management Program. The site is managed for DOE by Bechtel Jacobs. The site, approximately 13 miles west of downtown Oak Ridge, has nearly 500 facilities on about 2,200 acres and is the home of the Toxic Substance Control Act (TSCA) Incinerator. Primary remediation concerns include uranium and volatile-organic-contaminated groundwater plumes and surface water, solid low-level waste burial grounds, and decontamination and decommissioning (D&D) of uranium-contaminated buildings having about 15 million square feet (~344 acres) under roof.

Y-12 National Security Complex

The National Nuclear Security Administration's Y-12 National Security Complex is located in the Bear Creek Valley of East Tennessee immediately adjacent to the inhabited portion of Oak Ridge, Tennessee (population of 28,000), and about 15 miles from Knoxville. The site contains 811 acres, with some 500 buildings that house about 7 million square feet of laboratory, machining, dismantlement, and research and development (R&D) areas. The site boundary is 400 yards from the nearest Oak Ridge resident. The site is managed for DOE by B&W Y-12.

The complex was constructed as part of the World War II Manhattan Project. Construction began in February 1943, and operations began in November of that year. The first site mission was the separation of uranium-235 from natural uranium by the electromagnetic separation process. In the years following World War II, Y-12 evolved into a high-precision manufacturing, assembly, and inspection facility while maintaining the nation's uranium and lithium technology base. Y-12's missions have expanded since the end of the Cold War and the ensuing easing of international tensions.

The column exchange (COLEX) process that was operated at Y-12 to separate lithium isotopes used large quantities of elemental mercury. Lithium-6, separated from natural lithium by the COLEX process, was used to produce tritium for nuclear weapons. From 1950 to 1982 an estimated 2 million pounds of mercury at Y 12 were either lost to the environment or otherwise unaccounted for (EM Tour Book 2007).

The current Y-12 mission includes:

- Production and rework of complex nuclear weapon components;
- Receipt, storage, and protection of special nuclear materials;
- Quality evaluation and enhanced surveillance of the nation's nuclear weapon stockpile;
- Dismantlement and disposition of weapon components;
- Prevention of the spread of weapons of mass destruction; and
- Support to DOE, other federal agencies, and other national priorities.

Y-12 also applies its unique expertise, initially developed for highly specialized military purposes, to a wide range of manufacturing problems to support the capabilities of the U.S. industrial base. Y-12's all-inclusive expertise includes proceeding from concept, through detailed design and specification, to building prototypes and configuring integrated manufacturing processes.

ORNL

On February 2, 1943, ground was broken for Clinton Laboratories, or the X-10 site, as ORNL was then named. By summer, some 3,000 construction workers had erected about 150 buildings. The heart of the laboratory was an experimental reactor, a graphite cube 24 feet on each side with 7-foot-thick concrete walls for radiation shielding (far larger and more advanced than Fermi's Chicago pile) for converting uranium into plutonium. The small quantities of plutonium produced were used by chemical engineers to determine how to extract and purify it on a large scale. Besides supplying experimental quantities of plutonium to the California researchers, the Graphite Reactor and its chemical-separation labs served as pilot-scale models for Hanford's production plants.

ORNL's involvement with nuclear weapons ended after the war. During the 1950s and 1960s, ORNL became an international center for the study of nuclear energy, especially concerning the nuclear fuel cycle and waste management, and related research in the physical and life sciences. ORNL's nuclear involvements included development of the COLEX process, which resulted in mercury losses to the environment at the ORNL site, and the production of radioisotopes for beneficial use. With the creation of DOE in the 1970s, ORNL's mission broadened to include a variety of nonnuclear energy technologies and strategies. Today, ORNL is DOE's largest science and energy laboratory. ORNL has six major mission roles: neutron science, energy, high-performance computing, systems biology, materials science at the nanoscale, and national security. ORNL is managed for DOE by a partnership of Battelle and the University of Tennessee (UT-Battelle).

The ORNL site is approximately 10 miles southwest of downtown Oak Ridge and occupies about 2,900 acres. The ORNL site includes a variety of cleanup challenges: solid low-level waste burial grounds and pits, surface impoundments, Molten Salt Reactor D&D, Core Hole 8 groundwater plume, hydrofracture facility sites, gunite and associated tanks, mercury, buried transuranic waste, and degraded isotope production facilities.

ORR Cleanup Program Structure and Scope

The EM cleanup program for the ORR is managed by the DOE Oak Ridge Operations Office (ORO) and implemented by Bechtel Jacobs Company, LLC (BJC) and EnergX, which took over operation of the Transuranic Waste Processing Center in 2006.[2] An Advisory Board, composed of citizen volunteers, provides input on the Oak Ridge cleanup program. The goal of the program is to complete all cleanups within the scope of the program by 2015.

There are features of the ORR cleanup that make it particularly complex:

- With the two specific exceptions noted above, BJC performs cleanup activities at all three sites on the ORR while other site activities at Y-12 and ORNL are implemented by different contractors;
- ORNL has a number of facilities needing cleanup at the Y-12 site; and
- There are 439 facilities (about 5.3 million square feet of floor space) at ORNL and Y-12 needing cleanup that are not included in the scope of the current EM cleanup program. Of these facilities, 222 are not

[2] A request for proposals for transuranic waste stabilization at ORNL was issued as a Small Business Set Aside in FY 2009.

the responsibility of EM, but instead of the DOE Offices of Science and Nuclear Energy, and the National Nuclear Security Administration.

ORO has proposed a new Integrated Facility Disposition Project to address the facilities not included in the ongoing cleanup program. The project is estimated to cost $5 billion to $8 billion (unescalated dollars) and take 26 years depending on available funding.

CLEANUP TECHNOLOGY GAPS IDENTIFIED FOR THE ORR

Technology gaps presented in this section are based on presentations and discussions during the committee's March 2007 workshop and its site visit to Oak Ridge in June 2007. They are organized according to the program areas in the draft EM Roadmap presented to the committee in November 2007.

Program Area: Waste Processing

The EM roadmap's program area "waste processing" includes mainly high-level waste (HLW) issues. Despite the Oak Ridge site's early production of experimental amounts of plutonium, it never reprocessed nuclear fuels on the large scales of Hanford, Idaho, and Savannah River. Oak Ridge did not report any HLW-related issues at the workshop or during the committee's site visit.

Relevant to HLW tank cleaning at other sites, Oak Ridge completed cleaning eight concrete-walled (gunite) tanks in 2001. The tanks were constructed during the Manhattan Project in 1943 and used until the early 1970s. Importantly, Oak Ridge participants at the committee's March 2007 workshop reported that closeout of the gunite tanks was a good example of how innovative technologies could be used to tackle difficult cleanup problems. They reported that the cleanout was a test bed for over 100 technologies, including concrete scabbling, scraping, and robotics. The use of these technologies allowed the site to empty and grout the tanks more than a decade ahead of schedule. The gunite tank project was the first of its kind to be completed in the United States (NRC 2007).

All together 65 inactive tanks were closed from 1995 to 2007. One inactive tank, W-1A, is scheduled to be remediated in the 2009-2010 period. Twenty currently active tanks may become candidates for future cleaning and closure (Van Hoesen 2007). Oak Ridge does not consider these to be HLW tanks (EM Tour Book 2007).

While Oak Ridge reported no HLW issues, a summary presentation noted four R&D or technology needs that fall in the general area of waste characterization and processing (Van Hoesen 2007):

- Nondestructive analysis and examination technology for transuranic (TRU) wastes with high neutron activity—ORNL has more TRU wastes that require remote handling (RH-TRU) than any other site in the DOE complex,
- Processes for treating low-level liquid waste for disposal after shutdown of the existing centralized treatment capability,
- Disposition of wastes with no current path for disposition, and
- Mobile waste mixing and retrieval systems for small tanks.

Program Area: Groundwater and Soil Remediation

Oak Ridge faces significant challenges related to buried waste, subsurface contamination, and soil and groundwater remediation. There is buried waste at ORNL, Y-12, and ETTP. Contaminated surface water, groundwater, sediments, and soils were reported at ORNL, Y-12, ETTP, and, with the exception of contaminated groundwater, at some offsite locations (EM Tour Book 2007).

Mercury is a significant challenge. During production of nuclear weapon materials from 1950 to 1982 an estimated 2 million pounds of mercury at Y-12 were either lost to the environment or otherwise unaccounted for (EM Tour Book 2007). Some of the mercury has reached the East Fork Poplar Creek floodplain downstream of Y-12.

Mercury levels in the creek have been reduced to below drinking water limits. However, mercury remains a concern because of its concentration in fish and aquatic life is increasing. Y-12 continues to be a source of mercury (EM Tour Book 2007). DOE has installed a system that treats water from a spring at Y-12 to remove mercury as the water comes to the surface. This Big Spring Water Treatment System has halved mercury discharges from Y-12 from 8 to 4 kg per year (Munger 2007). However, the future release of contaminants to the groundwater and to surface waters during D&D is a concern. D&D actions such as turning off sump pumps and removing physical barriers (foundation slabs) can release contaminants. In addition D&D activities can alter the currently prevailing geology/hydrology and release pockets of contaminants (Phillips 2007).

During the committee's visit, Oak Ridge described significant science and technology (S&T) challenges related to mercury. Efforts to reduce concentrations of waterborne mercury in East Fork Poplar Creek at Y-12 may not reduce methylmercury in fish to safe levels. Additionally, some fish in White Oak Creek at ORNL exceed state and EPA fish-tissue concentration thresholds for mercury. There is need to identify the source and physical and chemical forms of mercury reaching the creeks, identify transport pathways and mechanisms, and design an effective treatment system (Phillips 2007). Although presenters identified these challenges, the committee received no information about studies that would address them.

In addition to mercury, there is extensive contamination of groundwater beneath the industrial areas of the Y-12 site by uranium and other metals, solvents, and some radioisotopes. Sources of contamination are not well known and the complex subsurface geology makes it difficult to identify flow paths (NRC 2007). Especially significant are some 40 million pounds of uranium buried in trenches on the site. Unless removed, this uranium will require perpetual monitoring (NRC 2007). Much of the uranium is pyrophoric, which complicates remediation.

DOE has used a continuous pump-and-treat system at the east end of Y-12 to keep an underground plume of carbon tetrachloride from spreading further. Water is pumped to the surface, treated, and then released into a nearby creek. This is a large plume that is evidently being fed from an underground source of the carbon tetrachloride. The treatment system has not eliminated the source or significantly reduced the concentration of carbon tetrachloride in the plume, but it has been effective in limiting the plume's offsite migration (Munger 2007).

According to a site presentation, for many of the plumes there are no technologies that can effectively remediate volatile organic compounds in the fractured bedrock. The groundwater is aerobic; therefore, reductive dechlorination is not feasible. Where there are no unacceptable risks and where contaminated groundwater will not migrate off the site at above drinking water limits, Oak Ridge is investigating the requirements to apply for Technical Impracticability (TI) waivers (Phillips 2007). However, according to the presentation, decisions that include long-duration monitored natural attenuation and/or TI waivers are difficult for regulators to accept (Phillips 2007). To support a TI waiver at ETTP, Oak Ridge listed a need for experts with experience in developing and demonstrating the rational for the TI of remediation in hydrogeological systems like that of the site (Phillips 2007).

Bechtel-Jacobs, DOE's cleanup contractor, recently completed the capping of 145 acres in Melton Valley where ORNL buried radioactive waste for over 40 years (B-J Tour Book 2007; and Van Hoesen 2007). Such caps and other engineered controls will require monitoring for decades (Phillips 2007).

Another area that may require remediation is associated with Core Hole 8. The name refers to an area of groundwater contamination located in the central portion of ORNL. The soil became contaminated through a leak from a broken pipe at the inlet to Tank W-1A, a tank containing highly radioactive TRU waste. A plume emanates from the contaminated soil and goes into First Creek (EM Tour Book 2007) in the center of the main ORNL site.

Since 1994, DOE has implemented various actions to minimize the release of contaminants to the creek. The interior of Tank W-1A has been cleaned, but the contaminated soil and groundwater remain. Remediation of the tank and contaminated soil and groundwater will be addressed under future Comprehensive Environmental Response, Compensation, and Liability Act actions starting in FY 2009 and to be completed by 2011 (EM Tour Book 2007). Such remediation will be challenging because of the combination of depth and high radiation levels.

Oak Ridge provided the following prioritized summary of its groundwater and soil remediation technology needs (Phillips 2007):

1. Support for a TI waiver for ETTP;
2. Pyrophoric materials (i.e., metallic uranium) in Trench 13 at Melton Valley;
3. Sources, transport, and treatment of mercury-contaminated water;
4. Reduction of mercury in fish via source treatment;
5. Evaluation of natural attenuation processes for treatment of groundwater plumes;
6. Release of subsurface contaminants during D&D work;
7. In situ treatment alternatives for mercury-contaminated soils;
8. Characterization of contamination sources under storm drains;
9. Phytoremediation for mercury remediation and monitoring of East Fork Poplar Creek;
10. In-situ remediation of pyrophoric materials at the Bear Creek Valley Burial Grounds;
11. Viability of large-scale treatment of mercury-contaminated water; and
12. Performance assessment, monitoring, and verification of technologies to support risk-based end states.

Program Area: D&D and Facility Engineering

The ORR contains hundreds of facilities that will eventually need to be deactivated and decommissioned (NRC 2007). Most of the 202 facilities at ORNL in the current EM baseline slated for demolition are reactors, laboratory facilities with hot cells, and their associated support facilities (EM Tour Book 2007). Some buildings are in poor structural condition, many contain worker hazards (e.g., high radiation fields >100 rads/hour and chemical and biological contamination), and some are located near occupied buildings or populated areas. In some facilities, equipment, piping, and duct work contain pyrophoric and other hazardous materials (e.g., mercury and lithium hydroxide). Many of these facilities have already been deactivated and do not have water or process waste lines. At present, EM is spending about $10 million per year for surveillance of facilities at the site and may have to spend additional monies just to make the facilities safe enough for workers to decommission (NRC 2007).

The removal of transite siding was presented to the committee as likely being the site's biggest D&D challenge (McCracken 2007). D&D work will involve removal of over a million square feet of these asbestos-containing panels. These transite panels are being handled and treated as nonfriable asbestos capable of becoming friable, and are therefore being removed manually one panel at a time. The current baseline removal method uses one to two laborers in a man-lift to manually remove the bolts holding the transite in place, lift the panel onto a saddle in the man-lift, and lower the panel to the ground; see Chapter 2, Figure 2.10.

According to DOE, if the panels could be pulled down with a grappling arm and allowed to fall to the ground, which is done with most other building materials, much time and cost could be saved, and the safety hazards associated with manually handling these heavy panels—many are high and difficult to reach—could be eliminated. One need cited by Oak Ridge is a study to review the regulatory drivers compelling the current baseline, including an evaluation of the science that went into the requirements to treat transite as potentially friable asbestos. Another alternative would be development and demonstration of an efficient, cost-effective, remotely operated tool for transite removal (Summary Sheet 2007).

The Alpha 4 facility at Y-12 is a 600,000-square-foot, transite-covered, structural steel and concrete facility with three floors and a subbasement. Alpha 4 was used until 1962 for a process to separate lithium-6 from lithium-7. The scope of the remediation project is to demolish the facility. Work includes eliminating classification concerns; gathering additional building characterization data to support preparation of a well-defined scope of work to allow the D&D to be subcontracted; completing hazardous materials abatement to remove asbestos, mercury, and solidified lithium compounds; deactivating utilities; and removing equipment (EM Tour Book 2007).

Beryllium was used extensively at the ORR and is present in numerous facilities that are slated to be demolished, especially at Y-12. It is a significant inhalation hazard, causing respiratory inflammation at low concentrations and permanent lung damage at high concentrations. Some workers are especially sensitive to beryllium. Current techniques for measuring beryllium in air involve sampling for laboratory analysis, which can take days to produce results. The site needs real-time, field-deployable beryllium monitors that provide accurate measurements at picogram levels. Such monitors do not exist at present, and their development is an important site technology need (NRC 2007).

The ORR has other technology needs to improve the safety and cost-effectiveness of D&D activities at the site (NRC 2007):

- Remote characterization technologies and technical approaches for cleanup of highly contaminated, deteriorated structures that have confined spaces or are otherwise unsafe for human entry, for example, the thousands of miles of piping in the gaseous diffusion facilities that contain uranium deposits. In addition, sensors are needed for making accurate, real-time measurements in extremely high radiation fields. There may be technologies outside of DOE that could be applied at the site.
- Decontamination technologies and tools, including cost-effective remote decontamination processes and robotics technologies; dry decontamination technologies that can be used to remove high levels of contamination with minimal secondary wastes; decision tools for determining optimal decontamination approaches; and technologies and approaches for removal of equipment containing high levels of radioactive and hazardous contamination.
- Demolition technologies and tools for understanding, predicting, and preventing the release of contaminants during facility demolition; technologies and approaches for real-time monitoring during facility demolition; and technologies and approaches for demolition of highly contaminated structures near operating facilities and populated areas.

The demolition of tall (>100 foot [30 meter]), highly contaminated off-gas stacks is a good example of the site's demolition challenges. These stacks are too contaminated internally and too close to operating facilities to be knocked down. Dismantling them brick-by-brick would be expensive and potentially hazardous to workers.

Program Area: DOE Spent Nuclear Fuel

Oak Ridge did not include DOE legacy fuel rod assemblies among its EM cleanup challenges. Presenters did describe challenges in removing salt containing uranium-235 remaining in a reactor used to test molten salt as an alternative reactor fuel.

The Molten Salt Reactor Experiment (MSRE) operated from 1965 to 1969. Following operation of the reactor, preparations were completed for long-term storage of the fuel, which was drained into two tanks. Beginning in 1987 and culminating in 1994, surveillance detected a migration of radioactivity from the tanks to other process lines.

In 1998 a Record of Decision was approved for removal of fuel and flush salts. Testing of fuel and salt removal equipment was completed in FY 2003. Processing of the initial flush salt tank was completed in June

2005. Processing of fuel drain tank 2 was started in December 2005, but operations were halted in May due to a fluorine release. Recovery activities are in progress. The MSRE is a high-priority project to be completed by 2011 (EM Tour Book 2007), but there do not appear to be any additional R&D needs.

Program Area: Challenging Materials

Challenging materials include legacy waste materials for which DOE has not defined a disposition pathway. Oak Ridge reported an inventory of about 140,000 curies of actinide isotopes (Pu, Am-241 and Am-243, Cm-244 and Cm-248, Bk-249, Cf-252, and Es-253) that were orphaned in Building 7920 of ORNL's Radiochemical Engineering Development Center (REDC) when the Office of Science ended support for heavy-element work. The orphaned inventory also includes some 340,000 curies of mixed activation and fission products (Michaels 2007).

The Building 7920 hot cell facilities began operation in 1966. There was support for heavy-element research from DOE and its predecessors until 2006. Starting in 2007 the hot cells housed an integrated spent fuel processing demonstration for the Global Nuclear Energy Partnership (GNEP) supported by the DOE Office of Nuclear Energy (Michaels 2007).

ORNL's High Flux Isotope Reactor (HFIR) produced much of the orphan actinide inventory. The HFIR itself uses beryllium neutron reflectors that become orphan waste when they are replaced. These reflectors are designated as TRU waste, but because they did not originate in a defense program they cannot be disposed in DOE's Waste Isolation Pilot Plant, which is designated for defense TRU waste.

LEVERAGING, COMMUNICATION, AND IMPLEMENTATION

In the mid-1990s some 30-40 percent of ORNL funding was from EM—roughly $208 million out of a total budget of $540 million. In 2007 EM funding made up only about 1.5 percent—about $15 million out of a total budget of $1,020 million (Michaels 2007). According to the presenter, the laboratory successfully transitioned away from EM, grew ORNL's science and technology business in other areas, and retained much of its nuclear processing and nuclear facility expertise (Michaels 2007).

There was little explicit discussion about how EM might leverage or utilize ORNL's nuclear expertise. Nonetheless, some potential opportunities were evident in other presentations. The laboratory overview noted that ORNL is "looking at nature in new ways" and is "working across scales to solve environmental challenges" (Roberto 2007). This new look at nature includes study of complex systems at multiple scales in space, time, and biological levels. The new look was also said to test and integrate ecosystems, observations and data, and advanced simulations (Roberto 2007).

A follow-up presentation (Jacobs 2007) provided an EM-relevant example of multiple-scale biogeochemistry with field-scale emphasis. This approach encompasses:

1. Fundamental science supported by the DOE Office of Science, including:
   - subsurface biogeochemical dynamics, and
   - uranium and an emerging mercury focus.
2. Applied R&D supported by EM and site funding, including:
   - biological monitoring of remediation performance, and
   - ecological restoration.
3. Related research not supported by DOE, including:
   - bioavailability of metals in soils,
   - groundwater plumes (chlorinated organics),
   - perchlorate fate, transport, and treatment,
   - threatened and endangered species, and
   - remediation site support at Dover AFB, Delaware, and Ft. Stewart, Georgia.

Non-DOE sponsors of the above research include the Strategic Environmental Research and Development Program and the Environmental Security Technology Certification Program of the U.S. Department of Defense (DOD), DOD installations, and industry (Jacobs 2007).

POTENTIALLY RELEVANT CAPABILITIES AND INFRASTRUCTURE

Field Research Center (FRC)

The FRC supports the DOE Office of Science's Environmental Remediation Sciences Program goal of understanding the complex physical, chemical, and biological properties of contaminated sites. In particular, the FRC promotes understanding of the processes that influence the transport and fate of subsurface contaminants, the effectiveness and long-term consequences of existing remediation options, and the development of improved remediation strategies. It includes a series of contaminated and uncontaminated sites in which investigators and students conduct field research or collect samples for laboratory analysis. FRC research also encourages the development of new and improved characterization and monitoring tools (FRC Brochure 2000).

The stated objectives of FRC research are to (1) quantify recharge pathways and other hydraulic drivers for groundwater flow and dilution of contaminants along flow pathways and determine how they change temporally and spatially during episodic events, seasonally, and long term; (2) determine the rates and mechanisms of coupled hydrological, geochemical, and microbiological processes that control the natural attenuation of contaminants in highly diverse subsurface environments and over scales ranging from molecules to watersheds; (3) explore strategies for enhancing the subsurface stability of immobilized metals and radionuclides; (4) understand the long-term impacts of geochemical and hydrologic heterogeneity on the remobilization of immobilized radionuclides; and (5) improve the ability to predict the long-term effectiveness of remedial activities and natural attenuation processes that control subsurface contaminant behavior across a variety of scales. The FRC was said to be vital to the ORR groundwater strategy and groundwater Record of Decision in 2015 (Phillips 2007).

Hot Cell Facilities

ORNL has built and operated some 36 nuclear hot cell buildings onsite. Their uses have included:

- Nuclear fuel reprocessing R&D,
- Nuclear reactor fuels and materials R&D,
- Radioisotope production and applications, and
- The Office of Science's heavy-element program.

Today ORNL is consolidating this work into four facilities, which were built prior to 1970. Any future EM work with highly radioactive wastes or other materials at Oak Ridge can be done in one of these facilities (Michaels 2007).

The REDC includes two facilities, Building 7920 and Building 7930. Building 7920 houses an inventory of orphaned actinide isotopes, which was described in the Challenging Materials section of this appendix. Building 7930, designed to be used as a pilot-scale reprocessing plant, was completed in 1967. This building includes seven hot cells. Three hot cells are essentially free of contamination, one is being used for high radiation and contamination work, and three are unused or used for storage.

The Irradiated Fuel Examination Lab (Building 3525) was constructed in 1963. This facility and General Electric's Vallecitos facility are the only two U.S. hot cell facilities capable of accepting full-length light-water reactor fuel for destructive examination. It was used for material packaging for EM from 1999 to 2003. Its current and future missions include GNEP support and the Nuclear Regulatory Commission's fuels test program.

The Irradiated Metals Examination Test Facility (Building 3025e), constructed in 1950 remains a state-of-the-art facility for metallurgical testing of irradiated metals. The facility includes six hot cells. It has a number of ongoing programs, none of which are supported by EM.

TSCA Incinerator

Oak Ridge's incinerator at the ETTP site is the only incinerator in the DOE complex permitted to burn chemical waste that is subject to TSCA or the Resource Conservation and Recovery Act as well as radioactive liquids and solids. Absent the incinerator, disposition of these wastes would be difficult for EM. The facility has operated since 1991 and has incinerated more than 30 million pounds of radioactive polychlorinated biphenyl (PCB) and hazardous wastes from the ORR and out-of-state DOE facilities. Although the incinerator was originally scheduled to be shut down at the end of September 2003, plans now are for the facility to remain open through FY 2009 to help EM meet the demand for treatment of low-level radioactive waste containing PCBs and other hazardous materials (EM Tour Book 2007).

REFERENCES

B-J Tour Book. 2007. Tour of DOE Oak Ridge Office. Environmental Management Cleanup Sites. Bechtel Jacobs Company LLC. April.

EM Tour Book. 2007. National Research Council Tour of Current and Completed EM Projects. Oak Ridge, TN: Department of Energy, Oak Ridge Operations Office, June 13.

FRC Brochure. 2000. Overview of the Oak Ridge Field Research Center. Handout to the Committee on Development and Implementation of a Cleanup Technology Roadmap. Oak Ridge, TN: Oak Ridge National Laboratory, June 13. Contact David Watson.

Jacobs, G.J. 2007. Contaminant fate and transport research at ORNL. Presentation to the Committee on Development and Implementation of a Cleanup Technology Roadmap, Oak Ridge National Laboratory, Oak Ridge, TN, June 13.

McCracken, S. 2007. Oak Ridge EM Science and Technology Plans and Challenges. Working lunch presentation to the Committee on Development and Implementation of a Cleanup Technology Roadmap, Oak Ridge National Laboratory,Oak Ridge, TN, June 14.

Michaels, G. 2007. ORNL Nuclear Hot Cell Facilities. Presentation to the Committee on Development and Implementation of a Cleanup Technology Roadmap, Oak Ridge National Laboratory, Oak Ridge, TN, June 13.

Munger, F. 2007. Nuclear cleanup showing promise. KnoxNews, June 4. Available at http://www.knoxnews.com.

NRC (National Research Council). 2007. Science and Technology Needs for DOE Site Cleanup: Workshop Summary. Washington, DC: The National Academies Press.

Phillips, E. 2007. Groundwater remediation on the Oak Ridge Reservation. Presentation to the Committee on Development and Implementation of a Cleanup Technology Roadmap, Oak Ridge National Laboratory, Oak Ridge, TN, June 13.

Roberto, J.B. 2007. Overview of Oak Ridge National Laboratory. Presentation to the Committee on Development and Implementation of a Cleanup Technology Roadmap, Oak Ridge National Laboratory, Oak Ridge, TN, June 13.

Summary Sheet. 2007. Oak Ridge Technology Summary Sheet. Improved Method for Transite Removal. Handout to the Committee on Development and Implementation of a Cleanup Technology Roadmap, Oak Ridge National Laboratory, Oak Ridge, TN, June 14.

Van Hoesen, S.D. 2007. EM status, Integrated Facility Disposition Project (IFDP), and R&D/technology needs. Presentation to the Committee on Development and Implementation of a Cleanup Technology Roadmap, Oak Ridge Nation al Laboratory, Oak Ridge, TN, June 13.

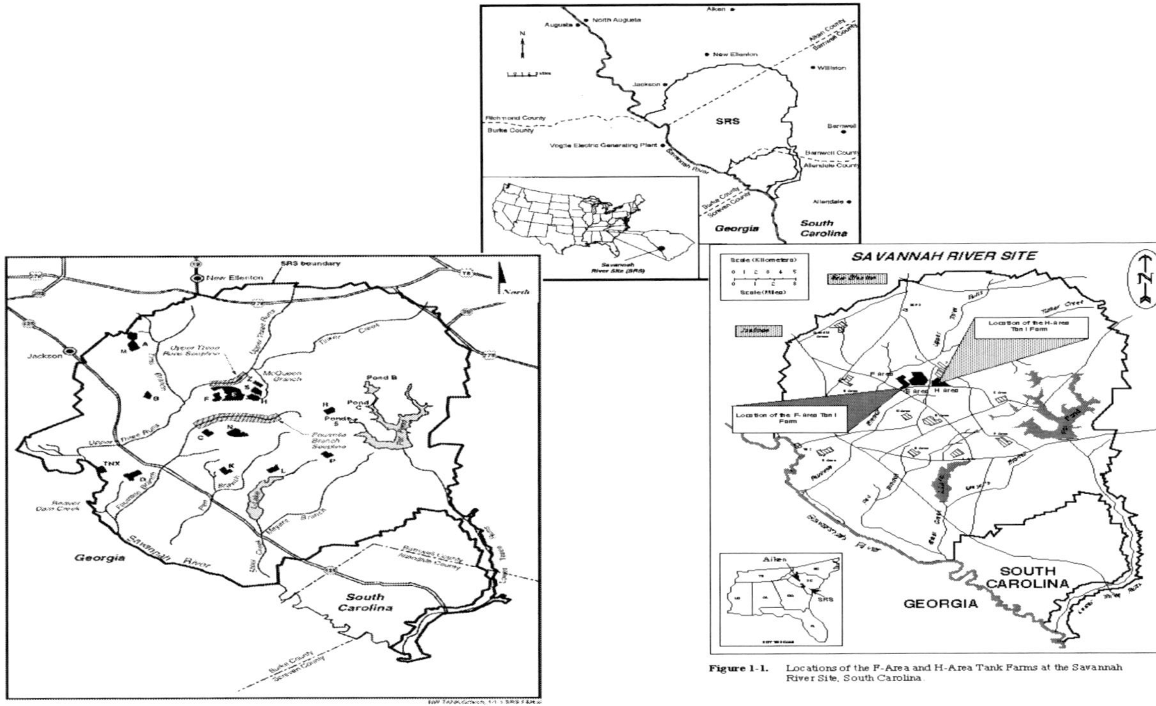

FIGURE G.1 The Savannah River Site occupies 310 square miles bordering the Savannah River in South Carolina. The site was established in the early 1950s when the U.S. government determined that there was a need for additional industrial-scale production of nuclear maerial in a location far away from the Hanford reservation
SOURCE: Department of Energy.

Appendix G

Savannah River Site

INTRODUCTION

The National Research Council Committee on Development and Implementation of a Cleanup Technology Roadmap held its fifth meeting in Augusta, Georgia on January 8-10, 2008. The purpose of the meeting was to obtain information relevant to the committee's Statement of Task (SOT) through presentations and tours by Department of Energy (DOE) staff and their contractors.[1]

This appendix provides a factual summary of the information related to the four items in the committee's SOT obtained during the meeting, the site visits, and documents provided to the committee. This appendix first describes the history and status of the DOE Savannah River Site (SRS), to provide perspective on the range of cleanup issues being managed by the DOE Office of Environmental Management (EM). The next sections summarize information presented to the committee, which guided the committee's deliberations in addressing its SOT as described in the main text. This appendix thus provides support for the findings and recommendations developed by the committee.

HISTORY

In 1950 E.I. duPont de Nemours and Company was asked by the then Atomic Energy Commission to design, construct, and manage a plant for

[1] The agenda for this meeting is shown in Appendix B.

producing nuclear weapons material on a 310-square-mile reservation bordering the Savannah River in South Carolina, see Figure G.1. DuPont had previously constructed the Hanford plant, and the U.S. government judged that a second production site, far away from Hanford, was necessary to ensure an adequate supply of weapons material as the United States entered the Cold War.

The Savannah River Plant (SRP) began producing heavy water for site reactors in 1952. R-Reactor, the first production reactor onsite, began operating ("went critical") in 1953. P-, L-, and K-reactors went critical in 1954, and the F-Canyon, a facility for reprocessing the reactor fuels, began radioactive operations. In 1955 C-Reactor and H-Canyon began operating, and the first plutonium shipment left the site. Construction of the basic plant was finished in 1956. The Savannah River Laboratory (SRL) provided research and development (R&D) capabilities to support the plant.

With the end of the Cold War, most of SRP's production activities were shut down between 1988 and 1992. In 1989 the Westinghouse Savannah River Company (which later became Washington Savannah River Company) became the site's primary management and operations contractor, and the site was designated as the SRS. In May 2004, the Secretary of Energy designated the former SRL as the Savannah River National Laboratory (SRNL). DOE selected Savannah River Nuclear Solutions, LLC, as the management and operations contractor for SRS in January 2008. In December 2008, DOE selected Savannah River Remediations, LLC, as the liquid waste disposition contractor at SRS. In addition to EM's site cleanup and environmental restoration work, SRS has a number of continuing missions, primarily dealing with plutonium and tritium processing.[2] The President's FY 2008 budget request for EM activities at SRS was about $1.4 billion (Allison 2008).

CLEANUP PROGRAMS AND CHALLENGES

This section describes ongoing cleanup programs and challenges as presented to the committee during its open meeting sessions at Augusta, Georgia, and its visits to SRS and to SRNL. The information is organized according to the program areas of the EM Science and Technology Roadmap (DOE 2008).

Program Area: Waste Processing

The SRS Waste Disposition Project addresses the site's liquid and solid wastes. Its liquid waste mission includes stabilization and disposition of

[2] See http://www.srs.gov/general/about/history1.htm.

36 million gallons of radioactive liquid waste currently stored in 49 underground storage tanks as a result of F- and H-Canyon operations, and closure of the emptied tanks.[3] The project also includes stabilization and disposition of transuranic, hazardous, mixed, low-level, and sanitary wastes (Spears 2008).

The 36 million gallons of tank waste at SRS contain nearly 400 million curies of radioactive materials, which is about half of the radioactivity in the DOE complex. The 27 tanks that have full secondary containment (i.e., they are double-walled tanks) are referred to as "compliant" tanks because they comply with the site's Federal Facility Agreement (FFA). The remaining 22 tanks that have partial or no secondary containment are referred to as noncompliant. The FFA requires that all noncompliant tanks be closed by FY 2022. Additionally, waste must be removed from all tanks by FY 2028 in accordance with the Site Treatment Plan and Consent Order (Spears 2008). Tank closure requirements are laid out in Section 3116(a) of the Ronald W. Reagan National Defense Authorization Act for FY 2005 (NRC 2006).

Most of the high-level tank waste originated from the reprocessing of nuclear fuels and irradiated targets for plutonium production via solvent extraction processes, which utilized nitric acid. The wastes were then made alkaline for compatibility with the site's carbon steel waste tanks. Most of the actinides and fission products in the waste, along with nonradioactive elements such as iron, precipitated from the alkaline mixture to form an insoluble sludge in the tanks. The remaining water-soluble salt solution, which contained some radioactive materials, notably Cs-137, was evaporated to conserve tank space. This resulted in a salt cake and supernatant solution that are also in the tanks. Spears (2008) reported that the SRS tanks contain about 16.9 million gallons of supernate, 16.6 million gallons of salt cake, and 3.0 million gallons of sludge.

Defense Waste Processing Facility (DWPF)

SRS began operating the DWPF in 1996. Its objective is to vitrify high-level tank waste to yield a stable form ready for disposal in a federal repository. Borosilicate glass developed at SRL was selected to be the waste form matrix in the late 1970s. To vitrify the waste, a glass-forming "frit" material is mixed with waste slurry and the mixture pumped onto the top of an already molten frit/waste mixture in a ceramic-lined melter ("slurry feeding"). The vitrification process takes place at about 1,150°C in the melter. Heat is provided by passing electricity directly through the molten glass (Joule heating). The glassmaking is a continuous process.

---

[3] SRS has closed 2 of its 51 high-level waste tanks.

Glass pours from the melter into a 10-foot tall, 2-foot diameter stainless steel canister. Filled canisters are allowed to cool, the top is welded shut, and the outside is decontaminated. The canisters are stored in a temporary onsite storage facility. At the time of the committee's visit, the DWPF had filled over 2,430 canisters with vitrified tank sludge (Spears 2008). Continued efforts to optimize the frit composition now allow about 38 weight percent of waste to be incorporated into the borosilicate glass waste forms (Davis 2008).

Salt Processing

Concurrent with design of the DWPF, which began in 1977, SRS designed an in-tank process intended to use an organic complexing agent (tetraphenyl borate) that could simply be added to a waste tank to precipitate the Cs-137. Supernate and dissolved salt cake would be pumped into a compliant tank designated for the process, the complexant added, mixed, and the insoluble Cs-tetraphenyl borate separated by filtration. This small, but highly radioactive Cs-137 stream would be vitrified in the DWPF along with the sludge. The large-volume, slightly radioactive "decontaminated" salt stream would be grouted into a product referred to as saltstone and emplaced in concrete vaults for permanent onsite storage (NRC 2000, 2006).

At about the same time that the DWPF began operating, the Defense Nuclear Safety Board and DOE determined that the in-tank precipitation process required further process chemistry assessment. During a test in Tank 48H, a substantial amount of flammable benzene from decomposition of the tetraphenyl borate was released into the tank (NRC 2000). Failure of the in-tank process left SRS without a means to disposition the bulk salts in its waste tanks and, hence, to empty the tanks. DWPF operations were essentially unaffected, since the tank sludge stream constitutes essentially all of the volume of waste that the DWPF was designed to vitrify.

Unable to remove any significant amount of waste from its tanks for the past 10 years, SRS has a severe shortage of tank space. Spears (2008) reported only about 1.3 million gallons of usable space remains after accounting for that needed for tank farm operations and contingencies. Ongoing operations, including waste recovery and DWPF operations will continue to consume space. SRS has, however, begun some salt processing on an interim basis.

Deliquification, Dissolution, and Adjustment (DDA)

Tank waste salts at SRS are predominantly sodium nitrate ($NaNO_3$), sodium nitrite ($NaNO_2$), and sodium hydroxide (NaOH), which result from rendering the nitric acid reprocessing waste alkaline. Evaporating this liquid waste to the extent possible to conserve tank space resulted in a mixture of supernate and salt cake, as noted previously. The salt cake is relatively depleted in Cs-137 because the hydrated radius of Cs^+ is larger than that of Na^+, so the cesium tends to be excluded from the crystallized salt. Conversely, Cs-137 is concentrated in the supernatant solution.

An expedient way to free tank space is to dissolve salt cake from selected tanks that contain relatively little Cs-137 and to send this material directly to the saltstone grout facility for permanent onsite disposal. At the time of the committee's visit, SRS was intending to implement this process, referred to as "deliquification (draining away the supernate), dissolution, and adjustment" (DDA) to process Tank 41 waste (Spears 2008). Because of entrainment of supernate and Cs-137 in the salt, DDA is not very effective for partitioning Cs-137, and its use will be limited to only dissolved salt cake from Tank 41.

Actinide Removal Process/Modular Caustic-Side Solvent Extraction Unit (ARP/MCU)

After failure of the in-tank precipitation process, SRS sought other options for salt processing and, after a detailed evaluation, selected a solvent extraction process tailored for alkaline waste. The process is based on a calixarene crown ether extractant (referred to as BobCalixC6), which is highly selective for cesium in the presence of sodium. Development of the extractant began with basic research at Oak Ridge National Laboratory (ORNL) and was subsequently supported by EM through its Environmental Management Science Program. To initiate this efficient means of partitioning Cs-137 from the salt as soon as possible, SRS designed and constructed a "modular caustic-side solvent extraction unit" (MCU). This is essentially a pilot-scale unit intended to help recover tank space and to fully demonstrate the solvent extraction process (Spears 2008).

SRS will also remove the traces of actinides (mainly plutonium) and Sr-90 that are in the salt waste using an ARP. In the ARP, a sorbent—monosodium titanate (MST)—is mixed with the salt solution, which is then filtered to remove the MST along with its adsorbed radionuclides. This filtered solution is sent to the MCU, and the MST is sent to the DWPF. Both the MCU and ARP were completing nonradioactive tests at the time of the committee's visit. Start-up testing with actual waste began later in 2008 (Spears 2008).

Salt Waste Processing Facility (SWPF)

SRS waste processing facilities operate under state-issued permits. The permitted limit for the total radioactivity to be disposed permanently onsite in the saltstone vaults is 1.4 million curies. Together the DDA and ARP/MCU processes might contribute 1.2 million curies (Spears 2008). SRS is therefore designing the SWPF. The SWPF is expected to process about 85 million gallons of supernate and dissolved salt cake at a rate of about 6 million gallons per year, while adding no more than about 0.2 million curies of radioactivity to the saltstone.

The SWPF will use the same processes as the ARP/MCU at a larger and more efficient scale. For example, the centrifugal contactors (high-speed rotating devices that mix and then separate aqueous and organic phases in the solvent extraction process) will be larger and there will be more extraction stages than in the MCU. Davis (2008) stated that an SRNL-modified monosodium titanate (MMST) formulation that shortens Sr-90 and actinide removal times in the ARP will play an important role. Deploying MMST in lieu of MST will increase salt processing throughput such that the risk of not meeting the Site Treatment Plan requirement to vitrify all current and future high-level waste (HLW) by 2029 is reduced. Construction of the SWPF had begun at the time of the committee's visit, and it is expected to be completed by the end of 2013 (Spears 2008).

Tank 48 Recovery

The in-tank precipitation test referred to previously left Tank 48H with about 240,000 gallons of highly radioactive liquid waste that also contains about 21,800 kilograms of organic compounds. Removing this waste will allow the 1.3-million-gallon compliant tank to be returned to tank farm service. SRS expects that Tank 48 will serve as a feed tank for the SWPF (Spears 2008).

The Tank 48 treatment process had not been selected at the time of the committee's visit. The process will be required to provide the capability to destroy the organics as well as to treat the salt waste. Options, which include fluidized bed steam reforming and wet air oxidation, are being developed by SRNL, Idaho National Laboratory (INL), and Pacific Northwest National Laboratory. Project completion is targeted for 1 year after SWPF startup to support maximum feed rates (Spears 2008).

Technology Needs

Spears (2008) named the following technology needs:

- Sludge heel removal, including improved chemical cleaning and mechanical cleaning;
- Sludge mass reduction, which essentially involves removing non-radioactive constituents from the sludge so that the amount of waste that must be vitrified is reduced;
- Waste processing, primarily increasing the production rate and loading of DWPF glass; and
- Ensuring the availability of methods to recover Tank 48.

Davis (2008) listed similar DWPF-related technology needs:

- Increase the waste loading in glass,
- Increase DWPF throughput, and
- Improve sludge preparation and qualification in the tank farm.

Davis (2008) noted that the cost of operating the HLW system is about $500 million per year. Salt processing and sludge vitrification are the rate-limiting steps; thus technologies that increase the rate of salt or sludge processing can reduce the life-cycle cost. Some options for increasing the melt rate include improving the glass-forming frit, improving the ability to mix the contents of the melter, and using a higher temperature alternative melter design. He also said that improving the sludge feed preparation steps in the tank farm can also increase DWPF throughput. These steps include removing aluminum from the sludge and washing the sludge to remove sodium salts—both sodium and aluminum increase the amount of glass required to vitrify a given amount of sludge. The rate of sludge settling after it is washed is the limiting step in these operations, and Davis (2008) suggested that the rotary microfilter is a new technology that could support continuous, rather than batch, sludge washing.

Davis (2008) stated that completion of salt processing is now expected to be about 2 to 5 years after sludge vitrification is completed. This is largely due to the salt processing difficulties described earlier in this appendix. One way to accelerate salt processing is to augment the SWPF with a process called small-column ion exchange. The process would include relatively small ion exchange columns mounted in risers (access ports) on the top of a compliant tank. The ion exchange media could be either an elutable, resorcinol formaldehyde organic resin or a nonelutable crystalline silico-titanate (CST) resin. The CST process would be simpler to operate but would produce more waste to be vitrified in the DWPF. Either resin would be fed from a rotary microfilter also mounted in a tank riser. Davis (2008) judged that the ion exchange technologies are maturing and could be deployed in about 3 years. He suggested that small-column ion exchange could avoid $2.5 billion in cost by reducing the DWPF life-cycle schedule by 5 years.

Lastly as a technical challenge, Davis (2008) stated that SRS tank waste processing must operate as an integrated system. All of its components, including the tank farms, SWPF, DWPF, and saltstone, must be kept operating simultaneously.

Tank Waste Retrieval and Tank Closure

Of the 24 noncompliant tanks, two have been closed and four more were expected to be ready for closure by the end of FY 2010 (Spears 2008). Tank closure requires removing waste from a tank until only an acceptable amount of residual waste remains. Once the tank is deemed "clean enough" it is filled with engineered grout. In the first two tanks to be closed, there are three grout layers: a chemically reducing grout at the bottom to maintain the radionuclides and toxic heavy metals in their most stable forms, a controlled low-strength material to fill most of the space, and then a stronger "cap" at the top to discourage intruders. Future closures may use chemically reducing grout to fill the entire tank (NRC 2006).

The use of carbon steel as the material for underground tanks necessitated neutralizing the tank contents rather than leaving them in their original acidic form. As a result, metal oxides and hydroxides precipitated, forming a sludge at the bottom of the tank that contains a high proportion of the radioactivity and is very difficult to remove. Oxalic acid is effective for chemical cleaning, but it causes degradation of the carbon steel (NRC 2006).

The two tanks that have been closed so far were selected from among the easier ones to clean, since they had no cooling coils; greater difficulties are anticipated with future cleaning campaigns (NRC 2006). Forty-three of the remaining compliant and noncompliant tanks have extensive cooling coil systems—some 20,000 to 25,000 feet of 2-inch-diameter carbon steel pipe per tank (Davis 2008). These coils present obstacles to the cleaning of the tanks by blocking water spray in their "shadows" and making it difficult for mechanical waste removal equipment to navigate around them. In addition the vertical pipes could represent "fast flowpaths" from the near-surface tank tops to residual waste on the tank bottom. These flowpaths must be eliminated, either by filling the pipes with stable material or cutting the pipes, before the tanks can be closed (Davis 2008).

The tank farm also has an extensive intertank waste transfer system. This includes 3-inch-diameter stainless steel pipes within carbon steel jackets or concrete encasements. Requirements for closing these transfer lines have not been finalized; however, Davis (2008) noted that approaches used at the INL site might be useful for SRS as well.

Program Area: Soil and Groundwater

SRS identified a total of 515 waste units—areas where there is contaminated soil, groundwater, and/or surface water—for remediation. The SRS Area Completion Project (ACP) is responsible for remediating these waste units, if warranted, as well as for facility deactivation and decommissioning. Whitaker (2008) reported that since 1993 the project has met all 1,990 of its scheduled FFA milestones and Resource Conservation and Recovery Act permit commitments, and 352 of the 515 waste units have been dispositioned. He added that the ACP's current approach addresses large groupings of waste units and facilities in a geographic area, rather than dealing with individual units.

SRS has 14 groundwater contamination areas. For remediation, each plume is considered to be in three parts:

1. The source area or "hot spot,"
2. The primary groundwater plume, and
3. The dilute plume, which leads the primary plume in the direction of the groundwater flow.

Hot spot remediation involves thorough characterization of the source and highly aggressive technologies, such as excavation, heating to drive off volatile compounds, in situ chemical oxidation, and active soil vapor extraction. For the primary plume, characterization and groundwater extraction are optimized to reduce the treatment volume. Technologies may include air stripping, recirculation wells, hydraulic barriers, phyto-irrigation, and base injection. For the dilute plume ahead of the primary, characterization is needed to predict mass transfer and flux. Remediation involves low-energy technologies such as passive soil vapor extraction, and natural attenuation. Currently 14 active groundwater remediation systems are operating to address the groundwater contamination areas (Whitaker 2008).

Whitaker (2008) described some of the more significant soil and groundwater program activities. One is a steam injection and contaminant removal system that is remediating a 3-acre area regarded as the primary source of subsurface contamination in A- and M-Areas. This dynamic underground stripping system is expected to complete the remediation in 5 years versus an estimated 200+ years using conventional technologies. The system had removed 380,000 pounds of solvents at the time of the committee's visit.

Electrical resistance heating removed 710 pounds of solvents at C-Reactor in 2006. The system achieved 99 percent efficiency according to soil samples, and completed the cleanup 2 years faster than soil vapor extraction. In 2007, the aboveground equipment was relocated to an area where chemicals, metals, and pesticides were disposed in a pit (referred to as the CMP Pit) (Whitaker 2008).

SRS consolidated the waste from three seepage basins into the area where low-level solid wastes were originally disposed, referred to at the "Old Radioactive Waste Burial Ground." A geosynthetic cover was then constructed over the 76-acre burial ground to close that facility. Phytoremediation is controlling tritiated groundwater, which originates in the area of the old burial ground and discharges to a stream. The control system includes a sheet pile dam to create a 2-acre pond, and using the pond water to irrigate 22 acres of pine forest that evapotranspires the tritiated water. This has reduced tritium entering the stream by 70 percent (Whitaker 2008).

Technology Challenges and Needs

Whitaker (2008) presented a prioritized list of technology challenges for the ACP, and he noted that the continued development and deployment of new technologies are critical to project success. The challenges he listed are the following:

1. Mass transfer limitations that affect removing contaminants from "tight zones,"
2. Remediating abandoned sewer lines,
3. Demonstrating monitored natural attenuation (MNA) and enhanced attenuation (EA) for chlorinated solvents,
4. Technologies that can support long-term institutional control, and
5. MNA and EA for metals and radionuclides.

Contaminants in tight zones include organics, metals, and radionuclides across the entire SRS due to its variably layered geology. Possible technologies to overcome mass transfer limitations in these zones include fracturing the clay to create openings, vadose heating to increase mass transfer rates, and long-term development of sustainable passive barriers (Whitaker 2008).

Abandoned sewer lines are responsible for diffuse, non-point source contamination. Innovative or improved tools are needed for characterization and remediation of up to 10 miles of underground lines associated with each industrial area at SRS. Possible technologies could include geophysics, gas tracers, robotics, in situ removal, and/or stabilization systems (Whitaker 2008).

For transitioning the site to long-term stewardship, monitoring tools will be necessary to demonstrate that natural attenuation is occurring as expected or to implement EA to ensure that continued active remediation is not required. Demonstrations of EA, for example, barometric pumping, long-lived electron-donor (e.g., vegetable oil) injection, and post-thermal treatments, are also needed. MNA/EA demonstrations will be needed for chlorinated solvents, metals, and radionuclides (Whitaker 2008). Technologies to support long-term stewardship and institutional control include improved tools and strategies (do we look at individual plumes or collectively at an entire watershed?), alternatives to current practice that require frequent measurements in large numbers of monitoring wells, and innovative systems for monitoring remediations that depend on caps or waste isolation (Whitaker 2008).

Program Area: Deactivation and Decommissioning (D&D)

The SRS ACP integrates facility D&D with soil and groundwater remediation. Whitaker (2008) reported that SRS had 985 total excess facilities and that 246 had been decommissioned at the time of the committee's visit. He described the T-Area completion as an example of a completed, integrated project. T-Area, formerly referred to as TNX, was an engineering semiworks area that used non-enriched uranium, but no other radioactive materials. Eight waste units were remediated, D&D was completed for 28 facilities, and a 10-acre geosynthetic cover installed. Groundwater remediation is under way (Whitaker 2008).

The reactor in P-Area is the first SRS reactor that will undergo D&D. Initial D&D work at P-Reactor has included removing the heavy-water moderator, friable asbestos, and historical artifacts; mold abatement; installing temporary power and lighting; and restarting building exhaust fans on temporary power (Allison 2008). Overall the P-Area project encompasses 100 acres and includes five waste units. Early characterization of tritium, solvents, and cesium contamination is complete. A Record of Decision for P-Area is scheduled for FY 2010 (Whitaker 2008).

Program Area: Spent Nuclear Fuels (SNF) and Nuclear Materials

Complexwide, DOE identified about 21 metric tons (MT) of surplus weapons-usable highly enriched uranium (HEU) and about 2 MT of surplus non-pit plutonium. This includes HEU in 19,500 SNF assemblies and 7.5 MT of HEU materials. In August 2006 the Deputy Secretary of Energy approved the continued operation of the H-Canyon at SRS as the preferred alternative for dispositioning these materials. H-Canyon is expected to remain in operation for this purpose until 2019 (Allison 2008). The SRS nuclear materials program includes surplus enriched uranium, SNF, and surplus non-pit plutonium (McGuire 2008).

Surplus Enriched Uranium

SRS is currently storing aluminum-, stainless steel-, and zirconium-clad fuels in L-Area. The site continues to receive fuel from both foreign and domestic research reactors. As part of the HEU disposition project, SRS will receive, in addition to its current inventory, aluminum-clad fuel from Idaho; in turn SRS will ship its stainless steel- and zirconium-clad fuels to Idaho for disposal. The aluminum-clad fuels will be dissolved and processed in H-Canyon to recover the HEU, which will then will be mixed with low-enriched uranium ("downblended") to provide uranium that is suitable for power generation but not usable for weapons (McGuire 2008). As noted earlier in this appendix, H-Canyon was designed and used for HEU processing to support the site's former weapons material production mission (NRC 1998).

Surplus Non-Pit Plutonium

SRS is storing significant quantities of plutonium materials that were produced at SRS and returned from the former Rocky Flats Site. DOE has also begun to consolidate surplus non-pit plutonium from Hanford, Los Alamos National Laboratory, and Livermore National Laboratory at SRS. Most of these materials have been stabilized and packaged in accordance with DOE Standard 3013 (NRC 2003). SRS will either disposition this material or repackage it for long-term storage (McGuire 2008). McGuire (2008) noted that dispositioning the material could involve H-Canyon, a new mixed-oxide (uranium and plutonium) fuel fabrication facility (MFFF), which is being built at SRS, and a proposed plutonium facility to prepare this material for processing at the MFFF.

Technology Needs

McGuire (2008) noted that there may be alternatives to the current plutonium disposition strategy, and that DOE seeks advice on technology needs to assist in evaluating potential alternatives. He also described technology needs associated with surveillance and storage of the DOE Standard 3013 containers. These include:

- Design and demonstration of furnace technology to oxidize and stabilize plutonium metal,
- Dustless material transfer technology (which would be an integral part of the furnace technology),
- Design of a modular sand filter that allows adding filtering capacity as needed (rather than building and operating a full-size sand filter from the

disposition project's outset—sand filters are a final step to ensure air that passes through a facility is safe to release to the environment), and

- A device to make welds on the outside of a 3013 can that comply with the standard.

CAPABILITIES AND INFRASTRUCTURE AT SRNL

The SRL was established in 1951 to provide R&D support for nuclear materials production at the SRP, now the SRS. In May 2004, the Secretary of Energy designated the laboratory as the Savannah River National Laboratory (SRNL). In early 2006, SRNL was further designated as EM's corporate laboratory. SRNL's total funding for FY 2007 was $154 million. DOE provided $139 million of this total, including $67 million from EM, $50 million from the National Nuclear Security Administration, $7 million from the Office of Science, and $15 million from other DOE offices (Marra 2008).

As the EM corporate laboratory, SRNL supports all EM closure activities at SRS and, in addition, assists and helps coordinate EM technology development and application at other sites. To explain the concept of a corporate laboratory, Gilbertson (2008) stated that EM is responsible for SRNL from an institutional perspective, and the laboratory serves as a resource for EM. Marra (2008) referred to SRNL as an "embedded" national laboratory. Marra (2008) noted that EM activities (e.g., waste management, environmental restoration) have been a major SRS mission for 40 years.

Marra (2008) listed SRNL's core capabilities as:

- Process development, pilot testing, design, and construction;
- Regulatory document and start-up support; and
- Production support and process optimization.

To provide these capabilities in EM areas, SRNL staff have expertise in:

- Radioactive chemical processing;
- Glass waste forms and vitrification process development;
- Application of environmental remediation technologies;
- Development and qualification of nuclear material packaging and nuclear fuel storage and handling; and
- Ultra-low-level, high-sensitivity nuclear measurements (SRNL 2007).

SRNL has a variety of both unique and traditional laboratory facilities for research and prototype development, including:

- Shielded cells—special containment facilities that provide the shielding and confinement necessary for examination, analysis, and testing of highly radioactive materials;
- Glove-box facilities—sealed, protectively lined compartments with attached gloves that allow workers to handle hazardous materials safely;
- Radiochemistry and analytical laboratories with contained instruments;
- Remote systems laboratory for the design, development, fabrication, and testing of equipment for use in radioactive, hazardous, or inaccessible environments;
- Engineering development laboratory for performing tests and demonstrations of equipment and existing or proposed designs;
- High-pressure test facility with steel-walled cells for high-pressure hydrogen exposure and testing, fatigue testing, and fracture toughness testing of metal specimens;
- Atmospheric Technologies Center with extensive capabilities for worldwide meteorological forecasts and real-time atmospheric transport modeling and assessment;
- Ultra low-level underground counting facility located 50 feet below ground that allows high-sensitivity measurements of ultra-low amounts of radioactivity;
- Advanced fracture mechanics laboratory with extensive capability for fracture testing in harsh environments and modeling to support system or component life extension;
- Primary standards laboratory providing calibration services compliant to the requirements of the American National Standard;
- Rapid fabrication facility, which produces low-cost prototypes, parts, and complete working models;
- Gamma irradiation facility for testing materials' abilities to withstand radiation exposure;
- Materials processing and fabrication laboratory to conduct materials processing, including powder metallurgy and solidification processing of nuclear materials; and
- Digital radiography facility that provides a highly sensitive alternative to traditional film x-rays for looking at the contents inside a container, verifying the quality of welds, and detecting deformations (SRNL 2007).

Shedrow (2008) noted that SRNL has developed a variety of technologies that have been applied to environmental remediation at SRS and other locations. These include:

- Optimized groundwater remediation systems,
- Field screening and technology deployments,

- Special sensors,
- Barrier monitoring and containment,
- Waste disposal forms,
- Wetlands remediation,
- Environmental biotechnology,
- Fate and transport modeling, and
- Environmental dosimetry.

Shedrow (2008) added that SRNL is the national lead laboratory for the monitored natural attenuation/enhanced attenuation project for chlorinated solvents. She also stated that SRNL has developed and applied innovative solutions for dense nonaqueous phase liquid characterization and remediation.

REFERENCES

Allison, J. 2008. Savannah River cleanup. Presented to the Committee on Development and Implementation of a Cleanup Technology Roadmap, Savannah River Site, Augusta, GA, January 8.

Davis, N. 2008. LWO technology development. Presented to the Committee on Development and Implementation of a Cleanup Technology Roadmap, Savannah River Site, Augusta, GA, January 10.

DOE (Department of Energy). 2008. Engineering and Technology Roadmap: Reducing the Uncertainty in the EM Program. Washington, DC: DOE.

Gilbertson, M. 2008. Comments to the committee. Presented to the Committee on Development and Implementation of a Cleanup Technology Roadmap, Savannah River Site, Augusta, GA, January 10.

Marra, J. 2008. An overview of the Savannah River National Laboratory. Presented to the Committee on Development and Implementation of a Cleanup Technology Roadmap, Savannah River Site, Augusta, GA, January 9.

McGuire, P.W. 2008. Nuclear Material Stabilization Project overview and technology needs. Presented to the Committee on Development and Implementation of a Cleanup Technology Roadmap, Savannah River Site, Augusta, GA, January 8.

NRC (National Research Council). 1998. Research Reactor Aluminum Spent Fuel: Treatment Options for Disposal. Washington, DC: The National Academies Press.

NRC. 2000. Alternatives for High-Level Waste Salt-Processing at the Savannah River Site. Washington, DC: The National Academies Press.

NRC. 2003. Improving the Scientific Basis for Managing DOE's Excess Nuclear Materials and Spent Nuclear Fuel. Washington, DC: The National Academies Press.

NRC. 2006. Tank Waste Retrieval, Processing, and On-Site Disposal at Three Department of Energy Sites: Final Report. Washington, DC: The National Academies Press.

Shedrow, D.M. 2008. Environmental science and biotechnology overview. Presented to the Committee on Development and Implementation of a Cleanup Technology Roadmap, Savannah River Site, Augusta, GA, January 9.

Spears, T.J. 2008. Savannah River Site Waste Disposition Project. Presented to the Committee on Development and Implementation of a Cleanup Technology Roadmap, Savannah River Site, Augusta, GA, January 8.

SRNL (Savannah River National Laboratory). 2007. Savannah River Laboratory Overview. Factsheet available from SRNL public affairs, Savannah River Site, Aiken, SC, May.

Whitaker, W. 2008. Area Completion Project: Soil and groundwater remediation and facilities deactivation and decommissioning. Presented to the Committee on Development and Implementation of a Cleanup Technology Roadmap, Savannah River Site, Augusta, GA, January 8.

Appendix H

Interim Report

THE NATIONAL ACADEMIES
Advisers to the Nation on Science, Engineering, and Medicine

Nuclear and Radiation Studies Board

500 Fifth Street, NW
Washington, DC 20001
Phone: 202 334-3066
Fax: 202 334-3077
www.nationalacademies.org

February 14, 2008

Mr. Mark Gilbertson
Deputy Assistant Secretary for Engineering and Technology
Office of Environmental Management
U.S. Department of Energy
Washington, DC 20585

Subject: Technical and Strategic Advice for the Department of Energy, Office of Environmental Management's Development of a Cleanup Technology Roadmap – Interim Report

Dear Mr. Gilbertson:

The fiscal year (FY) 2007 House Energy and Water Development Appropriations Report directed the Department of Energy (DOE) to prepare a technology roadmap that identifies technology gaps in the current DOE site cleanup program and a strategy, with funding proposals, to address them. At the request of DOE's Office of Environmental Management (EM), the National Research Council (NRC) empanelled a committee to assist DOE in developing the roadmap (Sidebar 1).[1] You requested the committee, as a part of its ongoing study, to provide an interim report to inform EM's deliberations on its FY 2009 plans for cleanup technology development. This interim report responds to your request.

Considering the limited time available to prepare this interim report at about the midpoint in its study, the committee chose to summarize its initial observations that bear on the importance of a strong EM-directed research and development (R&D) program to meet EM site cleanup challenges, and to underpin these observations with a few important examples of needs and opportunities for EM R&D. The committee's final report, to be issued in February 2009, will be developed in accordance with the full statement of task.

The committee began its study with a March 2007 workshop at which DOE site representatives, regulators, and citizens described cleanup challenges and technology needs (gaps) at DOE's four major cleanup sites: the Oak Ridge Reservation, Tennessee; the Idaho National Laboratory; the Hanford Reservation, Washington; and the Savannah River Site, South Carolina. Technology needs identified during the workshop as well as those identified by previous NRC committees are summarized in the workshop report (NRC, 2007). These needs generally fall into all five program areas in EM's draft roadmap listed in Sidebar 1. Some of the

[1] The committee's statement of task for this study is included as Attachment A and the committee roster is included as Attachment B.

Mr. Mark Gilbertson
February 14, 2008
Page 2

sites' higher-priority, longer-term needs are presented as examples in the second part of this interim report.

To date, the committee has visited three of the four above-named sites[2] to complete its assessment of technology gaps and priorities, and to understand the research capabilities and infrastructure at their national laboratories that are relevant to EM needs. In February, 2008 the committee is holding a three-day closed session for detailed discussions of this study. In spring 2008, the committee will hold an information-gathering meeting in Washington, D.C., with representatives of EM, other DOE offices and federal agencies, universities, and the private sector to better assess how EM might leverage its R&D with other programs. The committee may also request additional information from the DOE sites before it completes its final report.

The committee generally agrees with the five program areas for strategic R&D initiatives presented in EM's draft Cleanup Technology Roadmap. However, based on the information it has gathered, the committee observes that implementing the roadmap will require substantial and continuing federal support for medium- and long-term R&D for technologies focused on high-priority cleanup problems. As used in this report, short-, medium-, and long-term refer to time periods on the order of 1-5, 5-10, and >10 years, respectively.

Observations

(1) The complexity and enormity of EM's cleanup task require the results from a significant, ongoing R&D program so that EM can complete its cleanup mission safely, cost-effectively, and expeditiously.

The wide range of operations carried out by DOE (and its predecessor organizations) during the past 60+ years has resulted in hazardous and radioactive waste accumulation in tanks, soil, groundwater, and buildings. The sheer size of the cleanup in terms of numbers of facilities, land area, and contaminated subsurface and groundwater volume is enormous— amounting to an estimated life-cycle cost of over $235 billion.[3] Within this tremendous undertaking, there are thousands of individual tasks. Many of these tasks are complex and unique (for example,

---

[2]The committee visited the Savannah River Site and the Savannah River National Laboratory in early January 2008, when this interim report was in review. This report's reviewers are listed in Attachment C.

[3] In March 8, 2007, testimony before the House Committee on Appropriations, Subcommittee on Energy and Water Development, Assistant Secretary for Environmental Management James Rispoli reported that the estimated life-cycle cost for the DOE cleanup program had increased to about $235 billion owing to the addition of new projects as well as regulatory and *technology development problems* [ital. added] with current projects. DOE's fiscal 2009 budget request, which was released while this interim report was being prepared for printing, puts the potential cost of removing or remediating radioactive waste and other contamination at the sites between $265 billion to $305 billion, see (http://www.cfo.doe.gov/budget/09budget/Content/Volumes/Volume5.pdf).

Mr. Mark Gilbertson
February 14, 2008
Page 3

cleanout of high-level waste tanks, separation and processing of radioactive wastes, and cleanup of structures and groundwater plumes contaminated with radioactive and chemically hazardous materials). Within each specific task, the compositions of the wastes or the contaminants, or other factors (e.g., tank or building age and structure, site geology), often differ sufficiently to require the work to be customized to the situation.

Congress and DOE have provided substantial funding for EM's investments in scientific research and technology development since EM was created in 1989. However, this funding has varied substantially—rising from $184 million in FY 1990 to almost $410 million in FY 1995, followed by a decade-long slide to around $20 million per year recently (NRC, 2007).

Several previous NRC committees from which EM has sought advice have recognized the need for a strong science and technology base for site cleanup work. The 1997 report Building an Effective Environmental Management Science Program (EMSP) stated that "given the size, and scope and long-term nature of DOE's cleanup mission, the committee views the establishment of the EMSP as a prudent and urgent investment for the nation" (NRC, 1997, p. 12).

Sidebar 1

A Brief Description of the Draft EM Cleanup Technology Roadmap

The technology roadmapping process has been widely used as a planning tool in industry and government to match technology resources with desired product or process outputs. In the case of industry, these outputs are often products to meet certain commercialization needs. In *Vision 2020: The Lighting Technology Roadmap*, DOE used this technique in working with industry to align resources to meet new challenges in building lighting systems (DOE, 2007a).

The draft EM roadmap lists five program areas that are central to site cleanup:

1. Tank waste processing (including waste retrieval and tank closure),
2. Groundwater and soil remediation (including buried waste, flow path, and contaminant characterization),
3. Facility deactivation and decommissioning,
4. DOE spent nuclear fuel, and
5. Challenging materials (generally speaking, these are nuclear materials with no definite path to disposition).

Technical risks and uncertainties are listed in tabular format for each of these program areas. For example, within tank waste processing, the roadmap indicates that there are technical risks and uncertainties involving waste storage, waste retrieval, tank closure, waste pretreatment, and stabilization. Strategic initiatives to address each uncertainty are also listed.

Mr. Mark Gilbertson
February 14, 2008
Page 4

A later committee concluded that the "uniqueness and complexity of DOE's EQ [environmental quality] problems demand that the EQ R&D portfolio have a strong, if not dominant, long-term component" (NRC, 2001, p. 4).

In directing EM to prepare the Cleanup Technology Roadmap, the fiscal year 2007 House Energy and Water Development Appropriations Report stated support for EM technology development work and cited another previous NRC report, as follows:

> "EM technology development program funding has declined over the years, while at the same time, many technological challenges continue to face the program. For example, the National Research Council's 2005 report on Improving the Characterization and Treatment of Radioactive Wastes recommends that 'an improved capability for environmental monitoring would strengthen EM's plans to leave waste and contaminated media at DOE sites,' and, 'Monitoring systems at EM closure sites have been estimated to be some 25 years behind the state-of-art.' The Committee directs the increase to address the technology short-falls identified by this report."[4]

After visiting three of EM's major cleanup sites and witnessing both the cleanup accomplishments and the enormity of the remaining cleanup tasks—as well as potential new tasks to be added from other DOE offices in the future—the committee judges that existing knowledge and technologies are inadequate for EM to meet all of its cleanup responsibilities in a safe, timely, and cost-effective way. Meeting current and future challenges will require the results of an ongoing R&D program.

(2) By identifying the highest cost and/or risk aspects of the site cleanup program, the EM roadmap can be an important tool for guiding DOE headquarters investments in longer-term R&D to support efficient and safe cleanup.

The committee recognizes that large sums of money are being spent to clean up DOE sites. This includes short-term applied R&D activities supported to varying degrees by cleanup contractors. The committee is concerned that the medium- and long-term research component of EM's program has largely disappeared.

Need for longer-term R&D: EM carries out its site cleanup mission by issuing contracts, usually through its site offices, for specified cleanup tasks. EM's cleanup work is thus being

---

[4] House Report 109-474 to accompany H.R. 5427, Energy and Water Development Appropriations Bill, 2007. The Appropriations Committee recommended a $10 million increase over DOE's initial budget request.

Mr. Mark Gilbertson
February 14, 2008
Page 5

carried out by contractors at each site who have incentives to get the job done safely, on schedule, and within budget. As the contractors find it necessary, they may engage the national laboratories, universities, or other organizations to provide technical assistance. Accordingly, most funding for R&D tasks to support cleanup is presently provided by cleanup contractors for near-term technical support. However, given that cleanup contracts typically last from three to five years, contractors cannot be expected to provide sustained support for medium- and long-term R&D to meet EM's broader technology needs during the next approximately 30 years that the cleanup program is now expected to last.

Cleanup contractor-supported R&D is analogous to the industry practice of funding product-related R&D through business units. Experience from industry indicates that such units, driven by the profit/loss bottom line each quarter, make investments only for short-term results and incremental product improvements. Longer-term investments reduce the short-term financial performance of business units and are not generally funded by those units. However, it is the longer-term investments that are more likely to result in new product and process concepts. In industry, strategic R&D investments are usually made at the corporate level to ensure the future availability of innovative products.

Samsung, for example, describes its R&D funding in three tiers ranging from the business unit for product development, to division-level for core competencies, and corporate for future platform technologies.[5] By analogy, the role of DOE headquarters (corporate) would be to provide sufficient funding for integrated medium- to long-term R&D needs identified in collaboration with site cleanup contractors to support major improvements in the sites' cleanup operations.

Efficient approaches to addressing cleanup problems: Cleanup contractors typically bid on jobs according to a scope of work. However, EM cannot specify a scope of work or manage a contract effectively without first understanding the nature of the cleanup problem. For example, to scope a remediation task for the cleanup or containment of buried waste or a subsurface contaminant plume, a basic understanding of the problem would include the probable mechanisms and pathways by which contaminant migration could occur; how the contaminant migration could be stopped, curtailed, or intercepted; and the most effective remediation options that a contractor might implement. Such understanding of a cleanup problem is often based on the results of longer-term research, which as noted above, is seldom funded by the cleanup contractors.

The importance of research in understanding the nature of a cleanup problem was illustrated by Pacific Northwest National Laboratory (PNNL) in seven examples of apparently

[5] http://www.samsung.com/us/aboutsamsung/companyprofile/researchanddevelopment/CompanyProfile_RD_WorkforceOrganization.html.

Mr. Mark Gilbertson
February 14, 2008
Page 6

anomalous contaminant migration at Hanford—the contamination was moving in unexpected amounts and/or directions. The reasons underlying the apparently anomalous behavior were resolved in each case by scientific study that led to improved approaches for remediation or containment of the contamination (Stewart, 2007).

Developing alternatives to baseline approaches requires a similar understanding of the cleanup problem. Other NRC committees have concluded that most cleanup requirements within EM's current scope can be met, but new technologies can provide more technical options that may make the work more efficient and less risky (e.g., safer and/or more likely to meet performance and cost objectives). One example, which was mentioned frequently to the committee, was the development of a solvent extraction method for removing cesium from tank waste. The new method resulted from basic research followed by an EMSP grant for applying this research to an EM problem. After exploring several alternative technologies for high-level waste salt processing, the Savannah River Site is implementing solvent extraction for cesium removal (NRC, 2000, 2006).

Whereas near-term technology needs are recognized and generally fulfilled by the cleanup contractors through outreach to appropriate resources, support for medium- and long-term research and technology development requires a plan (i.e., technology roadmap) that identifies high-priority R&D needs and defines a program to meet these needs.

(3) The national laboratories at each site have special capabilities and infrastructure[6] in science and technology that are needed to address EM's longer-term site cleanup needs. The EM roadmap can help establish a more direct coupling of the national laboratories' capabilities and infrastructure with EM's needs.

Dating back to the Manhattan Project, R&D at national laboratories led to the nation's first nuclear weapons and weapons material production. National laboratories played key roles in supporting large-scale production of materials for nuclear weapons throughout the Cold War. They also built on this expertise by expanding into areas such as nuclear energy and beneficial uses of radioisotopes.[7] Although the missions of the national laboratories have expanded to include most areas of cutting-edge science, expertise in basic radiochemistry, radiochemical separations, remote equipment operation and maintenance, nuclear instrumentation, and radiation monitoring remains a forte and is essential to addressing EM cleanup challenges. The laboratories also retain production-era infrastructure, including shielded hot cells where substantial amounts of highly radioactive materials and wastes can be handled. State-of-the-art

[6] The statement of task directs the committee to identify the national laboratories' capabilities and infrastructure relevant to EM needs. As working definitions, the committee considers "capabilities" to refer to the expertise of laboratory personnel and "infrastructure" to refer to facilities and equipment.

[7] Nuclear energy and isotopes programs would seem to offer opportunities for leveraging EM investments with other DOE offices, although they have not yet been discussed by the committee. A previous NRC (2003) report suggested possible beneficial uses for EM's excess nuclear materials.

Mr. Mark Gilbertson
February 14, 2008
Page 7

computing facilities, which are part of today's national laboratory infrastructure, are also needed by EM, for example, to model cleanup options and estimate their effectiveness. New capabilities and infrastructure, such as those at the Oak Ridge Field Research Center, are clearly important for EM's work.

As production-era personnel retire from operations and the national laboratories, their knowledge of the former production facilities and waste disposal sites, which EM is tasked to clean up, will disappear unless there is sufficient EM support to attract new investigators with whom the experienced personnel can work to transfer their knowledge and expertise. Additionally, without EM support for university research, faculty will have little incentive to train the students who will provide future expertise for EM-related R&D.

As one would expect, the degree to which expertise and infrastructure are directed to cleanup problems is commensurate with the level of EM-headquarters and contractor support in the national laboratories' budgets. Relatively little EM work from either source is being supported at Oak Ridge National Laboratory (ORNL), which received about $15 million in total EM support in 2007 (Michaels, 2007). PNNL, which received about $91 million total from EM in 2007, provides substantial support for the Hanford cleanup (Walton, 2007). At PNNL most of the EM funds came through the cleanup contractors and were directed at site services (e.g., dosimetry), subject matter expertise, (e.g., tank waste chemistry or subsurface fate and transport), or near-term technology issues. Because of mission change, the Idaho National Laboratory (INL) has significantly shifted its research support of EM cleanup to short-term responses, although the laboratory has capabilities in many areas, especially in subsurface science that is necessary for the understanding of the fate of soil contaminants at each of the nuclear waste sites.

In 2006, DOE designated SRNL as the "corporate laboratory" for the DOE Office of Environmental Management.[8] In this capacity, SRNL has the responsibility to apply its unique expertise and technology capabilities to reduce technical uncertainties in meeting cleanup requirements across the DOE complex.

The EM roadmap can help establish a more direct coupling of national laboratory capabilities and infrastructure with EM's high-priority long- and medium-term R&D needs. The committee's final report will assess the national laboratories' capabilities and infrastructure that will be needed to address EM's long-term, high-risk cleanup challenges, and how their support might be leveraged with other programs at the laboratories.

[8] See http://srnl.doe.gov/newsroom/2006news/em-corp-lab.pdf.

Mr. Mark Gilbertson
February 14, 2008
Page 8

Needs and Opportunities for EM Research and Development

As stated at the beginning of this report, the committee generally agrees with the five program areas[9] listed in the draft EM roadmap. This section describes some of the higher-priority, medium- to long-term needs in the draft roadmap's program areas, and is based on the science and technology needs for EM cleanup discussed at the March 2007 workshop (NRC, 2007) and during the committee's three site visits.

Tank waste cleanup: A very expensive and long-term problem for the EM cleanup program involves retrieval of waste from the tanks at the Savannah River Site (SRS) and Hanford, processing the waste to separate the radionuclides into a high-level waste stream and a low-activity waste stream (intended to contain mostly non-radioactive chemicals), and converting these streams to monolithic solid waste forms destined for deep-underground or near-surface disposal, respectively. Tank waste retrieval and tank cleanup present challenges that are likely to be different for each tank. While some steps in the cleanup process can be used repeatedly in several tanks, the nature of the wastes, the configuration of the tanks, and a host of other factors dictate the process for cleanup of each tank. For example, the Hanford Office of River Protection presented a list of technology needs for its tank cleanup, and estimated that $109 million of R&D funding would be necessary to address these needs during the next 5 years (Mauss, 2007). A previous NRC (2006) committee examined challenges of tank waste cleanup at Hanford, INL, and SRS. Its final report described additional R&D needed to improve waste retrieval, waste processing, and tank closure.

Tank waste immobilization: Borosilicate glass was selected in the late 1970s as the baseline waste form for immobilizing tank sludge, primarily because of its long-term durability and its ability to incorporate a wide variety of waste constituents. However, use of borosilicate glass to immobilize DOE tank waste requires considerable pretreatment to remove bulky (e.g., sodium salts) and low-solubility (e.g., chromium) chemicals to increase the amount of waste that can be incorporated per volume of glass (waste loading).

EM has an important opportunity to develop alternatives to the borosilicate glass baseline for waste processing. Other waste forms may allow higher waste loadings and/or be fabricated more economically and faster, while meeting the anticipated requirements for disposal. Iron-phosphate-based glasses and metal matrixes were described to the committee as possible alternatives that may provide much higher loadings and better durability than borosilicate glass. The committee was also briefed on an induction heating method that might produce borosilicate or other glasses more efficiently and offer potentially significant advantages over Joule heating,

---

[9]The initial draft of the EM roadmap (DOE 2007b) included only the first three areas listed in Sidebar 1. Mark Gilbertson added the last two in a revision of the roadmap that he described to the committee at its Richland, Wash., meeting on November 2, 2007. The committee did not discuss needs and opportunities in these last two areas before drafting this interim report.

Mr. Mark Gilbertson
February 14, 2008
Page 9

which is the current baseline (Roach and Gombert, 2007). Such alternatives could provide large cost savings, since the cost of operating the processing and solidification facilities, such as those planned for Hanford, is at least $500 million annually (Mauss, 2007).

Groundwater and soil remediation: Subsurface contamination at the major sites includes inorganic materials such as uranium, technetium, and mercury as well as organic materials such as chlorinated solvents. Remediation requires characterization of subsurface contamination, understanding the soil structure and hydrologic conditions that will affect the mobility of the contaminants in the subsurface over long periods of time, technical options for remediation and/or containment of the contaminants, and an understanding of the longevity of containment options. Some important groundwater and soil remediation challenges remain unresolved at EM sites.

One ongoing challenge is the detection, removal, and/or containment of dense non-aqueous phase liquids (DNAPLs) such as carbon tetrachloride. Carbon tetrachloride occurs from near-surface to deep in the difficult-to-characterize fluvial gravels underlying the Hanford site, and it also occurs in fractured bedrock aquifers, including one of the fractured aquifers beneath the Oak Ridge Reservation. The complexity of remediating DNAPL contamination at the Oak Ridge Reservation's East Tennessee Technology Park is driving a request for a "technical impracticability" waiver from the State of Tennessee. Even if contaminant removal is precluded because cleanup is deemed technically impractical, science-based detection, monitoring, and decision-making protocols are needed to support arguments for such a technical impracticability waiver and ongoing risk management at the site. Basic understanding of how contaminant plumes may be attenuated by sorption, diffusion into low-permeability zones, biodegradation, and other processes can help EM determine the best approaches to deal with such contaminants.

Groundwater contaminants such as DNAPLs are also common problems at industrial sites. Although the committee has not specifically addressed its task item on leveraging, groundwater remediation is a likely opportunity for EM to leverage its work with private-sector organizations. Clearly EM is not working in isolation, and the leveraging would be expected to go both directions (i.e., industry can enlighten DOE and DOE can enlighten industry). Leveraging R&D with the Environmental Protection Agency, the federal regulator for hazardous chemical remediation, to strengthen the scientific basis for cleanup requirements would also benefit EM. The committee will discuss opportunities for leveraging in its final report.

Both Hanford and INL face complex challenges with the use of existing investigative approaches and technologies to monitor contaminant migration in the deep vadose zone. At Hanford there is also a need to develop effective and less costly remedial techniques for characterizing and managing or removing carbon tetrachloride and technetium-99 that are located in heterogeneous, partially or fully saturated sediments many tens of meters below the ground surface.

Mr. Mark Gilbertson
February 14, 2008
Page 10

Long-term performance of caps and barriers: EM is responsible for leaving sites in a condition suitable for long-term stewardship and is relying heavily on caps and barriers to contain buried wastes and contaminant plumes at many sites. Sustained R&D investments are needed to develop effective monitoring strategies for containment options. Ideally, such monitoring strategies would include sensor networks (external to and/or within barriers) that could provide real-time, long-term information (e.g., radiation levels, moisture, pH, temperature profiles) that is important to the cap and barrier performance. This information is needed sooner, rather than later, so that a realistic performance estimate at different sites, and under different conditions, can be constructed, and hence better predictive models can be developed to provide advanced warning of possible barrier failures as well as a knowledge base for further improvements in design and construction.

Facility deactivation and decommissioning (D&D): Transite panels were used as siding on many production-era DOE buildings. Removal of transite panels is an acute problem for decommissioning the gaseous diffusion plants at Oak Ridge (McCracken, 2007) and probably at other sites. Because production-era transite contains asbestos, worker health and safety regulations require careful handling to prevent breaking of pieces from the main panel, even though this siding is robust, non-powdery, and non-flaking. As a result of these regulations, workers have to manually handle the heavy transite panels often high off the ground and in a limited space (e.g., in a basket lift). According to Oak Ridge, the health and safety regulations applied generically to asbestos may actually increase the hazards to the workers who must remove these panels.[10] Improved science- and technology-based approaches might include the development of robotic devices to remove asbestos-bearing materials or a comprehensive risk assessment to provide a scientific basis for reviewing the relevant regulations.

Current plans for cleanup and closure of DOE sites often call for mid- to long-term stabilization of facilities awaiting future D&D or slated for long-term stewardship. Weathering and subsequent destabilization of these structures could result in release of contaminants to the environment. Retaining relevant expertise and supporting research programs to develop stabilization methodologies and technologies to limit the effects of building deterioration, while not hindering or complicating the building's future disposition, are important medium- to long-term challenges for EM. Maintaining aging buildings until they eventually undergo D&D will also require monitoring and sensing technologies, some of which could be leveraged from groundwater protection and remediation programs mentioned previously.

Conclusions

This interim report provides the committee's initial observations in its study to provide technical and strategic advice to assist DOE's development and implementation of the EM

---

[10] Oak Ridge Technology Summary Sheets: Improved Method for Transite Removal. Handout to the committee during its visit to the Oak Ridge Reservation, June 14, 2007.

Mr. Mark Gilbertson
February 14, 2008
Page 11

Cleanup Technology Roadmap. In concluding this interim report, the committee wishes to highlight the following:

- The committee generally agrees with the five program areas for strategic R&D presented in EM's draft Cleanup Technology Roadmap.
- According to the range of technology needs presented to the committee and the committee's initial observations, the committee judges that existing knowledge and technologies are inadequate for EM to meet all of its cleanup responsibilities in a safe, timely, and cost-effective way. Meeting current and future EM challenges will require the results of a significant, ongoing R&D program.
- The committee is concerned that the medium- and long-term research component of EM's program has largely disappeared. Implementing the roadmap will require substantial and continuing federal support for medium- and long-term R&D for technologies focused on high-priority cleanup problems.

The committee views the Cleanup Technology Roadmap as a continuing effort to establish an effective longer-term R&D program in support of EM's cleanup activities. The need for such a program has not diminished in the 11 years since the NRC (1997) report Building an Effective Environmental Management Science Program. Unless EM can provide substantial and continuing support for medium- and long-term R&D, its efforts to bridge current technology gaps, maintain needed capabilities and infrastructures at national laboratories, and initiate leveraging of other research programs are not likely to be effective.

Our final report, which will fully address the statement of task, will be completed in early 2009 in accord with the schedule we discussed with you during the committee's Richland, Wash. meeting.

Sincerely,

Edwin Przybylowicz

Edwin Przybylowicz, Chair

Allen G. Croff

Allen Croff, Vice Chair

Attachment A: Statement of Task
Attachment B: Committee Roster
Attachment C: Reviewers
Attachment D: References

Mr. Mark Gilbertson
February 14, 2008
Page 12

ATTACHMENT A
STATEMENT OF TASK

A National Academies committee will provide technical and strategic advice to the DOE-EM's Office of Engineering and Technology to support the development and implementation of its cleanup technology roadmap. Specifically, the study will identify:

- Principal science and technology gaps and their priorities for the cleanup program based on previous National Academies reports, updated and extended to reflect current site conditions and EM priorities and input from key external groups, such as the Nuclear Regulatory Commission, Defense Nuclear Facilities Safety Board, Environmental Protection Agency, and state regulatory agencies.
- Strategic opportunities to leverage research and development from other DOE programs (e.g., in the Office of Science, Office of Civilian Radioactive Waste Management, and the National Nuclear Security Administration), other federal agencies (e.g., Department of Defense, Environmental Protection Agency), universities, and the private sector.
- Core capabilities at the national laboratories that will be needed to address EM's long-term, high-risk cleanup challenges, especially at the four laboratories located at the large DOE sites (Idaho National Laboratory, Oak Ridge National Laboratory, Pacific Northwest National Laboratory, and Savannah River National Laboratory).
- The infrastructure at these national laboratories and at EM sites that should be maintained to support research, development, and bench and pilot scale demonstrations of technologies for the EM cleanup program, especially in radiochemistry.

The committee will provide findings and recommendations, as appropriate, to EM on maintenance of core capabilities and infrastructure at national laboratories and EM sites to address its long-term, high-risk cleanup challenges.

Mr. Mark Gilbertson
February 14, 2008
Page 13

ATTACHMENT B
COMMITTEE ROSTER

COMMITTEE ON DEVELOPMENT AND IMPLEMENTATION OF A CLEANUP TECHNOLOGY ROADMAP

EDWIN P. PRZYBYLOWICZ, chair, Eastman Kodak Company (retired), Webster, New York
ALLEN G. CROFF, vice-chair, Oak Ridge National Lab (retired), St Augustine, Florida
RICHELLE M. ALLEN-KING, University of Buffalo, New York
SUE B. CLARK, Washington State University, Pullman
PATRICIA J. CULLIGAN, Columbia University, New York City, New York
RACHEL J. DETWILER, Braun Intertec Corporation, Minneapolis, Minnesota
THOMAS F. GESELL, Idaho State University, Pocatello
GARY HALADA, State University of New York, Stony Brook
CAROLYN L. HUNTOON, CLH Associates, Inc., Barrington, Rhode Island
EDWARD LAHODA, Westinghouse Science and Technology Center, Pittsburgh, Pennsylvania
ROBIN ROGERS, University of Alabama, Tuscaloosa; The Queen's University of Belfast, Northern Ireland, UK
GARY S. SAYLER, University of Tennessee, Knoxville
ANDREW M. SESSLER, Lawrence Berkeley National Laboratory (retired), Berkeley, California
J. LESLIE SMITH, University of British Columbia, Vancouver, Canada

Staff
JOHN R. WILEY, Senior Program Officer
MANDI M. BOYKIN, Senior Program Assistant

Mr. Mark Gilbertson
February 14, 2008
Page 14

ATTACHMENT C
REVIEWERS

This report has been reviewed in draft form by persons chosen for their diverse perspectives and technical expertise in accordance with procedures approved by the National Research Council's Report Review Committee. The purposes of this review are to provide candid and critical comments that will assist the institution in making the published report as sound as possible and to ensure that the report meets institutional standards of objectivity, evidence, and responsiveness to the study charge. The review comments and draft manuscript remain confidential to protect the integrity of the deliberative process. We wish to thank the following for their participation in the review of this report:

Robert J. Budnitz, Lawrence Berkeley National Laboratory, Berkeley, California
Ken Czerwinski, University of Nevada, Las Vegas
Tissa Illangasekare, Colorado School of Mines, Golden
Milton Levenson, Bechtel International (retired), Menlo Park, California
Paul Locke, Johns Hopkins Bloomberg School of Public Health, Baltimore, Maryland
Walter L. Robb, GE Corporate Research and Development Center (retired), Schenectady, New York
Raymond Wymer, Oak Ridge National Laboratory (retired), Oak Ridge, Tennessee

Although the reviewers listed above have provided many constructive comments and suggestions, they were not asked to endorse, nor did they see the final draft of the report before its release. The review of this report was overseen by Chris G. Whipple, ENVIRON International Corporation. Appointed by the Division on Earth and Life Sciences, he was responsible for making certain that an independent examination of this report was carried out in accordance with institutional procedures and that all review comments were carefully considered. Responsibility for the final content of this report rests entirely with the authoring committee and the National Research Council.

Mr. Mark Gilbertson
February 14, 2008
Page 15

REFERENCES

DOE (U.S. Department of Energy). 2007a. Vision 2020: The Lighting Technology Roadmap. Washington, D.C.: DOE Office of Building Technology, State and Community Programs, Energy Efficiency and Renewable Energy. http://www.eere.energy.gov/buildings/info/documents/pdfs/lighting_roadmap_compressed.pdf.

DOE. 2007b. Engineering and Technology Roadmap: Reducing the Uncertainty in the EM Program. Washington, D.C.: DOE Office of Environmental Management. April. (Draft).

Mauss, B. 2007. Hanford HLW Program: Challenges to Cleanup. Presentation to the Committee on Development and Implementation of a Cleanup Technology Roadmap. Richland, Wash.: U.S. Department of Energy Office of River Protection, October 31.

McCracken, S. 2007. EM Science and Technology Challenges at the Oak Ridge Reservation. Presentation to the Committee on Development and Implementation of a Cleanup Technology Roadmap. Oak Ridge, Tenn.: Oak Ridge National Laboratory Operations Office, June 13.

Michaels, G. 2007. ORNL Nuclear Hot Cell Facilities. Presentation to the Committee on Development and Implementation of a Cleanup Technology Roadmap. Oak Ridge, Tenn.: Oak Ridge National Laboratory, June 13.

NRC (National Research Council). 1997. Building an Effective Environmental Management Science Program: Final Assessment. Washington, D.C.: National Academy Press. http://books.nap.edu/catalog.php?record_id=5557.

NRC. 2000. Alternatives for High-Level Waste Salt-Processing at the Savannah River Site. Washington, D.C.: National Academy Press. http://books.nap.edu/catalog.php?record_id=9959.

NRC. 2001. A Strategic Vision for Department of Energy Environmental Quality Research and Development. Washington, D.C.: National Academy Press. http://books.nap.edu/catalog.php?record_id=10207.

NRC. 2003. Improving the Scientific Basis for Managing DOE's Excess Nuclear Materials and Spent Nuclear Fuel. Washington, D.C.: National Academies Press. http://www.nap.edu/catalog.php?record_id=10684.

NRC. 2005. Improving the Characterization and Treatment of Radioactive Wastes for the Department of Energy's Accelerated Site Cleanup Program Washington, D.C.: National Academies Press. http://books.nap.edu/catalog.php?record_id=11200.

NRC. 2006. Tank Waste Retrieval, Processing, and On-Site Disposal at Three Department of Energy Sites. Washington, D.C.: National Academies Press. http://www.nap.edu/catalog.php?record_id=11618.

Mr. Mark Gilbertson
February 14, 2008
Page 16

NRC. 2007. Science and Technology Needs for DOE Site Cleanup: Workshop Summary. Washington, D.C.: National Academies Press. http://www.nap.edu/catalog.php?record_id=11932.

Roach, J. and D. Gombert. 2007. Presentation to the Committee on Development and Implementation of a Cleanup Technology Roadmap. Idaho Falls, Idaho: Idaho National Laboratory, August 29.

Stewart, T. 2007. Challenges to Cleanup: Soils and Groundwater Contamination. Presentation to the Committee on Development and Implementation of a Cleanup Technology Roadmap. Richland, Wash. Pacific Northwest National Laboratory, October 31.

Walton, T. 2007. PNNL 2007 EM Business Summary. Provided to the Committee on Development and Implementation of a Cleanup Technology Roadmap. Richland, Wash. Pacific Northwest National Laboratory, October 31.